The National Energy Problem

Edited by Robert H. Connery and
Robert S. Gilmour

Published for The Academy of Political Science

Lexington Books

D.C. Heath and Company
Lexington, Massachusetts
Toronto   London

Published simultaneously in Canada.

Printed in the United States of America.

International Standard Book Number: 0-669-93252-3

Library of Congress Catalog Card Number: 74-684

Contents

Some of the papers in *The National Energy Problem* were discussed at a conference sponsored by the Academy of Political Science on July 19, 1973, at Columbia University. The Academy serves as a forum for the dissemination of informed opinion on public questions, but it makes no recommendations on political issues. The views expressed in this publication do not necessarily reflect those of the Academy, the editors, or the government and academic institutions with which the authors are affiliated.

Contributors

MARVIN J. CETRON is the founder and president of Forecasting International and editor-in-chief of the *Technology Assessment Journal.*

VARY T. COATES is senior staff scientist and head of the Technology Assessment Group Program of Policy Studies in Science and Technology, the George Washington University.

ROBERT H. CONNERY is professor of government in Columbia University and president of the Academy of Political Science.

JAMES E. CONNOR is director of the Office of Planning and Analysis, U.S. Atomic Energy Commission.

HERBERT D. DRECHSLER is associate professor of mineral economics in the Henry Krumb School of Mines, Columbia University, and former vice-president of the American Metal Climax of New York.

ROBERT S. GILMOUR is assistant professor of government in Columbia University.

RICHARD L. GORDON is professor of mineral economics, College of Earth and Mineral Science, Pennsylvania State University. He is the author of a forthcoming book on the United States coal market.

ALLEN L. HAMMOND is research news editor of *Science,* the journal of the American Association for the Advancement of Science. He is coauthor of *Energy and the Future.*

CHARLES ISSAWI, an expert on economic policy and the Middle East, is Ragnar Nurkse Professor of Economics in Columbia University. He is coauthor of *The Economics of Middle Eastern Oil* and the editor of *The Economic History of the Middle East, 1800-1914.*

GENE P. MORRELL, a geologist and a lawyer, is vice president of the Lone Star Gas Company and cochairman of the National Petroleum Council Committee on Energy. He has served as director of the Office of Oil and Gas, U.S. Department of the Interior.

BARRY L. NICHOLS is project manager of the National Environmental Studies Project of the Atomic Industrial Forum. He has been an environmental consultant for government and industrial studies and has written extensively on the effects of nuclear energy on the environment and the potential of waste heat.

ROBERT P. OUELLETTE is director of environmental studies for the Mitre Corporation.

BORIS S. PUSHKAREV is vice-president for research and planning of the Regional Plan Association in New York City. He is coauthor of *Man-Made America: Chaos or Control?*, which received a National Book Award in 1964.

WINSTON W. RIDDICK, former director of the Institute of Government Research at Louisiana State University, is executive assistant state superintendent of education, Louisiana State Department of Education.

LAWRENCE ROCKS is codirector of the Energy Information Center, C.W. Post College. He is coauthor, with Richard P. Runyon, of *The Energy Crisis*.

RICHARD P. RUNYON, former dean of the Division of Science at C.W. Post College, is codirector of the college's Energy Information Center.

FRANCIS W. SARGENT is governor of Massachusetts and chairman of the National Governors' Conference Executive Committee on Transportation, Technology, and Commerce.

WILLIAM D. SMITH, petroleum editor of the *New York Times*, has written extensively on the national energy problem with special emphasis on oil.

JOAN EDELMAN SPERO is assistant professor of political science in Columbia University. Her major interest is national security.

CHARLES F. STEWART is professor of international business, Graduate School of Business, Columbia University, and a consultant to several oil companies on Middle East affairs.

Preface

It would be a mistake to view the United States energy problem only in terms of the present crisis. Resolving the crisis will not solve the long-term problem, which is the result of an increased demand for energy during the past decade. Consequently, Americans face a number of choices, none of which are pleasant for a society that has been accustomed to a prodigious and wasteful use of energy.

There would have been an energy problem during the winter of 1973-74 even without the Arab embargo. During the summer of 1973, electric brownouts and gasoline shortages clearly indicated future problems. Nevertheless, many Americans still do not believe that the nation has an energy problem of major proportions. Some think that it is simply a plot by the big oil companies to force the smaller ones out of business or to attain a higher price for their products. Others speculate that it is a ploy by the national administration to deflect attention from Watergate. However, the contributors to this volume agree that the United States does face a serious energy problem that will affect American life-styles and military strategy for many years to come.

The problem can be attacked either by increasing supply or decreasing demand. Demand can be reduced by using less heat in homes and offices and restricting the use of automobiles. Unfortunately, in many parts of the country, people do not have access to public transit. Reducing supplies to industry and transportation, the major consumers of energy, would cause unemployment. All of these factors make an equitable program for reducing demand difficult to develop.

Supplies of energy can be increased by the construction of new electrical power plants and by more extensive use of coal, solar energy, and atomic power. This increase will not only be expensive, but will also take time and require considerable improvement in engineering techniques. It may also adversely affect the environment. It is possible to increase supplies, particularly of petroleum, by importing supplies from foreign countries. But the United States must compete with the rapidly increasing demands of Western Europe and Japan. Consequently, the cost of imported energy has risen rapidly and undoubtedly will continue to become more expensive. Even if foreign sources are available, the quantities of petroleum that would have to be imported in the next decade would threaten the nation's balance of payments as well as the national security.

Probably there is no easy solution to America's energy problem either by substantially reducing demand or by increasing supply. A variety of means are available, however, that over time would help to solve the difficulties of the nation. *The National Energy Problem* not only spells out these difficulties in detail, particularly as they concern the next decade, but offers some specific programs for solving the problem. What is most needed is a national policy that clearly defines goals and establishes administrative procedures to meet them. Technical skills can be developed, new sources of energy can be found, and the environment can be protected, if all levels of government cooperate. To be sure, the president has submitted various proposals to Congress that are steps in the right direction, but much more is needed. The regulatory authority over energy facilities is still dispersed among several hundred federal, state, and local agencies. Until the efforts of these bodies are effectively coordinated to implement an agreed-upon national policy, little progress can be expected.

This volume is not intended for technical specialists but rather for the general reader. It attempts to bring together discussions of the major issues to be considered in formulating a national energy policy.

ROBERT H. CONNERY
ROBERT S. GILMOUR

The Nature of the Problem

The Energy Crisis

RICHARD P. RUNYON
LAWRENCE ROCKS

The mass media's proclamation of an energy crisis in the United States has evoked a widespread sense of dismay that an energy shortfall could have developed without anyone's being aware of it. The crisis exists nonetheless. The sporadic shortages during the winter of 1972-73 were mere portents of a chronic, severe, and perhaps irreversible energy shortage. Although the catastrophe will not unfold fully for approximately ten years, it will take at least that much time to develop and implement solutions. Environmental concerns and demands for energy are clearly on a collision course. The various and legitimate strategies for controlling and abating pollution have made increased demands upon energy resources. To complicate matters, both oil and natural-gas production in the United States, which provide a major portion of the nation's energy, are not sufficient for its needs.

Some energy experts proclaim the crisis a temporary problem brought about by the sudden environmental conscience that precipitated the 1969 Environmental Protection Act and the 1970 Clean Air Act. They point to the delays in the construction of the Alaska pipeline, the long court disputes over siting of power plants, the energy-profligate antiemission valve in the automobile, and the organized opposition to nuclear energy. Others blame the utilities for encouraging greater uses of energy faster than they were able to supply electricity and natural gas. Still others blame the crisis on a conspiracy of the major oil companies seeking to crush opposition to offshore drilling and the Alaska pipeline and to drive up the prices of gasoline and heating oil by creating a false shortage. Despite these views, the energy crisis is real and may be demonstrated by an objective appraisal of the facts.

The entire crisis can be summed up in three words, exponential growth rates. The population has been growing exponentially; of all the people who ever lived, approximately one-third are alive today. Because energy consumption per person has also grown exponentially, total energy consumption now outpaces the nation's capacity to produce sufficient energy from domestic sources.

Man is a 100-watt machine. In his primitive state he depended almost exclusively on these 100 self-generated watts for survival. Muscle and brain power were directed to exigencies of day-by-day existence. With the advent of agriculture, he began to harness the power of domesticated animals to supplement his own limited energies. Eons later, during the Industrial Revolution, man erected a vast technology to do his work.

Technologies, however, extract exorbitant energy demands. In the United States the per capita use of energy has nearly doubled in the past three decades while total energy consumption has tripled. Moreover, the per capita energy consumption in the United States is enormous compared to the rest of the world—10,000 watts to 1,200 watts. In other words, an American citizen uses about eight times as much energy as other persons on the planet. This prodigious demand for energy coupled with dwindling supplies of the most accessible and most utilized energy resources, natural gas and petroleum products, is the crux of the energy crisis in America.

There is, furthemore, a high positive relationship between per capita energy expenditure and the per capita gross national product.[1] The relationship between energy and material living standards indicates that solutions to the energy crisis must not be allowed to lead to a deep economic depression with all the personal, social, and political upheavals of massive unemployment.

The Nature of the Energy Crisis

The heart of the energy crisis is the great disparity between the rapid rate of oil and natural-gas depletion and the protracted rates at which any other sources can be phased in to replace them. The United States's

---

[1] Lawrence Rocks and Richard P. Runyon, *The Energy Crisis* (New York: Crown Publishers, 1973), p. 10.

production of oil and natural gas is now at its peak. Henceforth, these resources will be discovered and recovered at a diminishing rate, for the most part in the next two decades. The phasing in of other energy sources, however, will take many decades. For example, atomic power has two fundamental drawbacks. The supply of uranium 235 according to the most optimistic estimates is only equal to the total energy content of domestic oil. The breeder reactor can create plutonium fuel from inert uranium 238, but the breeding rate would only double the fuel inventory in ten years. This means that the breeder reactor could not possibly supply even one-half of the nation's present electrical needs until well after the year 2000. Furthermore, atomic energy is delivered as electricity, which cannot replace gasoline or oil in transportation, space heating, and power plants.

Geothermal energy is another potential source. Like atomic energy, however, geothermal energy is obtained as electricity. Optimists see a potential of 100 billion watts of electricity from geothermal sources. At present the United States uses 400 billion watts of electricity. Besides being insufficient, geothermal energy is also problematic. The heat-flow characteristics of geothermal dry sources have been determined by theoretical computer runs done at Los Alamos Scientific Laboratories. It will take years of research and development to ascertain whether the 100 billion watts of presumed geothermal electricity can be developed. Thus geothermal activity is a highly romanticized energy source with a protracted phase-in time.

Solar power has similar drawbacks. While it can indeed supply power to the nation, it cannot be phased in before the depletion of oil and gas, and it would also be delivered as electricity. Placing solar stations in space requires a space-shuttle program, which will take $5 billion and at least a decade to complete. Unless the cost of silicon cells comes down by a factor of 500, the prices of electricity generated in outer space will be preposterously high, at least 500 times the present cost. If solar stations are put on the ground, the sunshine gathered by day will be deployed during the night. Thus, a secondary fuel capacity —the so-called hydrogen fuel economy—is needed. To gather enough energy to electrify the power-grid would require the capture of about 10,000 square miles of sunshine. Building the necessary equipment, either parabolic mirrors and steam turbines or silicon cells, would require decades. Furthemore, hydrogen cannot be used as a fuel in cars, buildings, and power plants. It will take several generations of Americans to build systems to capture and deploy solar power. Meanwhile, this generation faces depleting domestic oil and gas reserves.

Domestic Issues

Shortages of natural gas and petroleum products have already produced modest price increases in energy. As the United States participates more and more in the international energy markets, it will have to compete directly with Japan and the European Economic Community for energy supplies. For political, economic, and technological reasons, it is highly unlikely that the Organization of Petroleum Exporting Countries (OPEC) will be able to increase the supply sufficiently to meet the demand. In the sellers' market that is rapidly developing, Americans may anticipate that oil and natural gas will be auctioned off to the highest bidder at prices impossible to estimate at this time.

Inflation will probably affect all energy-intensive activities in the American economy—agriculture, goods, and services (except communication). Together these activities amount to about 60 percent of the GNP. Increased oil and natural-gas prices would greatly increase production costs for these activties. Agriculture, for example, is almost totally mechanized. Energy is required to run the variety of farm machines that plant the crops, tend them during the growing season, and harvest them. Many of these corps must then be dried prior to processing. A substantial portion of the sugar-cane crop was lost during the winter of 1972-73 because shortages of natural gas forced the facilities that dry the harvest to be shut down. Finally, farm products must be shipped to market via carriers that are also energy-intensive. Clearly, increased energy costs will raise agricultural prices. A similar scenario can be written for all energy-intensive industries.

The facts of the energy crisis will demand that Americans carefully weigh all energy costs. Conservation of energy will become a major theme of national policy and will cause major industrial and commercial realignments. There will be movement toward more intracity and intercity mass transit and toward smaller, fuel-efficient automobiles. There may be lower suburban real estate values and repercussions on savings banks holding first mortgages, because homes in remote suburbs will become an economic liability.

Several energy-intensive industries may leave the country. The fertilizer industry has already left because of the more plentiful supplies of methane in Europe. The steel industry may be drawn away by Saudia Arabia's claims of producing steel at half-price by using gas supplies normally flared off at the wellhead. When a nation becomes energy-poor, it does not merely experience a shortness of breath—it runs out of heartbeat.

There is a strong temptation to lay all energy ills at the feet of the environmentalists. It is true that they strongly oppose virtually every energy procurement and deployment option, such as offshore drilling, the Alaska pipeline, deep-sea ports, supertankers, siting of power plants and refineries, strip mining for coal, and development of the breeder reactor. The Environmental Protection Act of 1969 gave teeth to the environmental movement, providing the legal machinery to delay various energy options. Moreover, the environmental movement appears to be ensnarled in many contradictions. Some environmentalists urge antipollution devices on automobiles but support the trans-Canada pipeline, which would be two to four times longer and almost certainly more destructive to tundra and forests than its Alaskan counterpart.

The environmental movement, however, is not to blame for the energy crisis. Oil and natural gas are running low because these finite resources are being used faster than they can be supplied. Man will continue to reach farther out to sea, deeper into the ocean floor, and farther into the Arctic Circle with a diminishing return for his efforts. It is clear, however, that the environmental movement has exacerbated the present energy shortfall and has accelerated the timetable for future shortages.

But to blame the environmental movement for the energy crisis is to reveal a myopic view of the past, present, and future. The environmental movement reveals the greatness in man, his concern for the quality rather than the quantity of life, and his recognition that he is both heir of past generations and trustee for future generations. Energy has been wasted without regard to the future. There has not been the necessary funding of research, neither scientific nor technological, to develop one or more of the eternal energy sources.

International Issues

To mention Middle East oil is to evoke conflicting and contradictory images. Many analysts see the danger of depending on the Middle East with its explosive political climate. Others see the possibility that the nations of OPEC will unite to blackmail the energy-dependent nations into modifying their positions on Israel and into paying increasingly exorbitant prices for oil and natural gas. What many energy analysts fail to perceive is that OPEC cannot supply enough oil even if these dire events do not happen. The rising rate of oil consumption in Japan, the European Economic Community, and the United States will require the importation of 20 billion barrels a year by 1985. OPEC produces only 9 billion barrels a year, and 40 percent of that is frozen for political and geological reasons. The remaining 60 percent capacity would have to be increased fourfold by 1985 to meet projected export demand. Such a rapid increase in oil exploration and production is held unlikely by most oil observers.

Moreover, the United States balance-of-trade deficit for energy alone could amount to between $30 billion and $40 billion a year by 1985. It is unlikely that the United States's exports to OPEC could cover this sum. Nor is it likely that these nations could consume the volume of imports necessary to effect a trade balance. Already Saudi Arabia, which has reached its saturation point for income, is putting extra oil income into world money markets. The Saudis must look hard to find places for additional investments. Hence they have little financial incentive for further exploration of their oil resources. It is more likely that excess dollars will be a source of mischief on the international money markets or come back in the form of investment in United States industry.

Another point, often overlooked, is that it is not in the best interest of OPEC to expand production to meet world demands. As a fuel, oil has a life span measured in decades. When the OPEC nations run out of oil, they are out of business. As a petrochemical, however, oil has a life span of at least 1,000 years. The OPEC nations would be well advised to put to work the billions of dollars that will come into their treasuries to develop indigenous petrochemical and allied industries. Such industries would create an educated middle class and sophisticated infrastructure. The secretary general of OPEC, Dr. Abderrahman Khene, has recently endorsed this view:

> The only way to deal with such a vital resource, taking in account the interests of the forthcoming generations, is to begin now to curb the trend of consumption of petroleum and gas as sources of energy. Indeed, it seems to me that it would be wise to replace, at least partially, the quantities to be saved by other sources, namely by coal, the reserves of which are almost unlimited, and possibly by cleaner sources of energy, as solar power for example. . . .
>
> OPEC countries will welcome such new trends for the use of petroleum and gas, in that way securing for the forthcoming generations a long-term period of important income, and a better possibility of access, on a large-

scale basis, to advanced technology which can assure them a higher standard of living.[2]

The depletion of oil resources will also affect relations with other areas. Although Canada remains a friendly neighbor to the north, there are growing strains over both energy and economic matters. Many citizens of the United States look hopefully to Canada's presumed reserves of 20 billion to 80 billion barrels of oil as an ace in the hole. However, it should be recognized that Canada is energy-rich only in relation to its small population. Massive exports of oil to the United States would rapidly deplete these energy reserves. Donald S. McDonald, minister of energy, mines, and resources, opposes too great an energy outflow to the United States. John Turner, minister of finance, has expressed deep concern about growing United States economic control over Canadian finances—two out of every three dollars in Canada are presently controlled by the United States. As recently as June 28, 1973, Prime Minister Pierre Trudeau warned, "While Canada is self-sufficient in terms of energy production, it does not have the potential to play a major role with regard to total North American needs."[3]

A third area of international activity involves the Soviet Union. It is estimated that the USSR has approximately four times America's reserves of oil and three times its natural-gas reserves. Because the Soviet Union's rate of utilization of these resources is low relative to population, it is expected that it will not be seriously affected by the energy shortage for several decades at least. In the meantime, the Soviets have offered to sell natural gas and enriched uranium to the European Economic Community (EEC), natural gas and oil to Japan, and gas to the United States in return for investment capital and technological talent to develop their oil and gas fields. Simultaneously, the USSR is in the process of expanding its capacity to refine Arab oil and serve as middleman in the oil markets, knowing that the United States has a retarded refining capacity because of environmental opposition to the siting of refineries.

In summary, the United States will soon become estranged from Canada over energy, will see its trading partners, Japan and the EEC, gravitate to the USSR for energy, and will fall into conflict with these countries over Arab oil, which is insufficient and will become too costly.

[2] Khene to Rocks and Runyon, April 20, 1973, files of Rocks and Runyon, Greenvale, New York.
[3] *New York Times*, June 29, 1973.

Solutions

The facts of energy require consideration of both the present and the future. Immediate steps must be taken to obtain sufficient energy to avoid an economic collapse within a decade, to build an energy bridge to the next century, and to find an eternal energy source, or a combination of several sources, to ensure future energy availability. Since the lead times required to phase in the energy options vary from years to decades, it is absolutely essential to take these steps now.

Several courses of action seem to be necessary: energy conservation, development of synthetic fuels from coal, and massive, immediate funding of research for the development of an additional long-lived energy source, such as fusion, sunlight, wind, and geothermal.

Energy conservation must become a major concern for these times and perhaps for the remainder of man's life on earth. Historically, energy has been used without regard to environmental consequences and the finite nature of the supply. Two-ton vehicles have been built to transport 150-pound persons, and buildings have been constructed without regard to heating and air-conditioning efficiency. Energy-profligate carriers, such as jet planes, have been used instead of energy-efficient carriers, such as river barges and railroads.

Space heating and transportation consume almost 45 percent of the nation's energy budget. If all ground transportation in the United States were conducted by railroad, the energy savings would be almost 20 percent of the nation's total needs.[4] Obviously, such a massive shift from automobile and truck to train is unlikely. If it were to occur, the economic and social dislocations resulting from a widespread abandonment of the automobile would be awesome. Nevertheless, enormous savings are possible by diverting some of the Highway Trust Fund to mass transit and by replacing the automobile emission standards of the Clean Air Act of 1970 with efficiency provisions. For example, by moving to fuel-efficient automobiles (smaller ones that necessitate fewer power options), Americans could save 1.5 billion barrels of oil a year by 1985. That would be 15 percent of the country's anticipated oil needs.

Similar energy savings could be effected in space heating. To cite one example, freezer and refrigerator units in most supermarkets vent their waste heat to the outside. By making minor structural changes, this

---

[4] Rocks and Runyon, p. 131.

heat could be circulated through the market, thereby alleviating the need for a separate heating plant.

Positive and negative incentives are the key to the success of conservation. By revising the rate structure of utilities to increase the cost per unit of energy consumed, rather than the reverse that now prevails, major users would discover that they can get along with less energy. Positive incentives can also be provided for individuals and businesses desiring to make structural changes in buildings for purposes of energy conservation. At the municipal level, a home should not be reassessed when the owner makes structural improvements to conserve energy. At the federal level, energy-conserving innovations should qualify as tax deductions for the individual home owner and as operational costs for the businessman.

As important as energy conservation is, it is not a final solution to the energy crisis. It will take many years to phase in the various energy-conserving options. For example, the 110 million automobiles on American highways cannot be replaced overnight with energy-efficient automobiles. Such an accomplishment would take about ten years after passage of automobile-efficiency legislation. Similarly, all the homes and buildings in the United States will not suddenly and magically require less heating. Certainly, given the best of incenitves, total energy conservation in space heating can be achieved only in decades. Meanwhile, the United States population continues to increase, and reserves of oil and natural gas continue to decline.

The energy bridge to an atomic- or solar-powered future appears to be coal, the giant of the fossil fuels. It can be directly burned or chemically converted into both a liquid and a gaseous fuel. When employed as a synthetic fuel, its potential life span may be measured in centuries. Moreover, the processes for synthesizing gas and oil from coal will require no scientific breakthroughs. Any one of several methods could be used by the end of the 1970s. There are environmental and economic problems that have to be solved before a synthetic-fuel industry can be developed. The enormous quantities of coal that will be needed can be obtained quickly and inexpensively only through strip mining. Without proper safeguards to assure restoration of the land, huge tracts of beautiful terrain west of the Mississippi will be desecrated. However, if strip-mined land is restored (at a cost of about $1,000 an acre), the dangers of economic collapse can be avoided and the surface use preserved for future generations.

The economic problem is, in many ways, more difficult to solve. If the United States presently had the capacity to produce synthetic fuels in commercial quantities, the cost would be too high to compete successfully with imported oil and natural gas. However, before the decade is over, the cost of these natural resources will unquestionably soar. Thus, before the end of the 1970s, synthetic fuels will surely possess a competitive advantage. The economic dilemma can be stated succinctly: How can the best chance of energy survival be phased in when initially it will not be economically competitive? It is clear that the United States must develop a financial strategy that will permit the synthetic-fuel industry to grow during a time when it will be expensive. A possible solution is price averaging, that is, the consumer pays for gas and oil in accordance with the industry's average costs, whether he uses natural or synthetic fuel.

Coal is a bridge, not a crutch. Coal could be of great value to future generations as a source of petrochemicals and high-grade proteins. If it is burned as a fuel now, this resource will be forever denied to following generations. To preserve America's coal heritage, the nation must be willing to make an immediate scientific and financial commitment to developing a long-lived and nonpolluting energy source, even though it may take as long as fifty years.

To date, most money for energy research has been spent on nuclear fission, particularly on the burner and breeder reactors. Fuel for the burner is too short-lived, and the breeder represents an environmentalist's nightmare. The greatest long-term hopes appear to reside in the fusion process, which has not yet been achieved in a controlled reaction. Fusion power would be essentially nonpolluting and eternal.

Some observers see solar power as the solution. And it is a fact that every day the sun sheds on the earth a power equivalent to 30,000 times present power needs. Others point to wind power, sea-thermal power, and geothermal power as major sources of future energy. No one will ever know which source or combination of sources will be the best unless an effort is made to develop them. In this respect, it is encouraging that President Nixon had advocated the expenditure of $2 billion a year over the next five years to support energy research.

However, science and technology alone cannot solve the energy problem. If the nation's institutions are to survive the power-base transformations that will take place during the next five decades, talent must be pooled from all areas of the physical and social sciences, as well as the oft-ignored humanities and the arts.

Energy in the New York Region

BORIS S. PUSHKAREV

Dramatic advances in material well-being throughout large parts of the world over the past three centuries were made possible by a vast expansion in the use of inanimate energy. Rising affluence, a result of mechanizing industry and agriculture, is in turn demanding more energy for comfort, transportation, and leisure. Moreover, it is demanding energy in cleaner and costlier forms, such as electricity and gas. Nowhere is this process more sharply pronounced than in the United States, which with 5.6 percent of the world's population is consuming one-third of the world's energy.

The fact that total energy consumption in the United States doubles every three decades and electricity every decade has been sufficiently publicized. Should energy use continue to increase at this rate during the next three decades, the amount consumed during that period will equal the total consumed in the United States since colonial times. And in the next decade, as much electricity would be needed as was generated in the United States since the Pearl Street plant was opened in New York City in 1882. By the end of the century, for every electricity-generating station now in existence, eight new ones of equal size would have to be built.

Thus it should not come as a surprise that the United States, a country that has always had cheap and abundant energy, is now confronted with some limits to unbridled growth in its use. First, the domestic supply of some preferred fuels, notably oil and natural gas, is lagging behind consumption. Alaska holds only a two-year supply at current consumption levels. Second, importing large quantities of fuel from abroad poses political and economic problems in the international arena. Third, there are purely logistic problems such as delays in the construction of complex nuclear plants and lagging research and development in the field of synthetic fuels from coal. Finally, the growing ecological efforts to curb the pollution caused by energy production and use also will be a limitation. Some efforts to clean up the environment are resulting in greater energy consumption in the short run (until more efficient ways of getting rid of automobile emissions are developed). Other efforts may perhaps be energy-intensive in the long run (recycling of materials). To make matters worse, the overall efficiency of energy conversion in the United States, which increased nearly fourfold in the first half of this century as a result of technological improvements, seems to be levelling off, in part because of inherent thermodynamic limitations.

Under these conditions, it is becoming increasingly clear that expanded supply is no longer the only tenable answer to expanded demand. Clearly, a large-scale research-and-development effort to implement new forms of environmentally compatible energy conversion must be coordinated with an effort to reduce the exponential growth in demand. The latter effort may prove more difficult in the short run, because it conflicts with many ingrained interests and habits. One might add that a number of countries with a comfortable living standard and a developed industrial base, such as West Germany or Sweden, have half the per capita energy consumption of the United States. So conservation measures do not require a return to the stone age. Rather, they only require curbing some peculiarly American extravagances.

In 1972, to explore opportunities for energy conservation, the Regional Plan Association, Inc., of New York and Resources for the Future, Inc., of Washington, D.C., undertook a joint study of energy consumption in the New York region. Sponsored by the Ford Foundation, the study focused on microscale differences in consumption that tend to be obscured in the usual macroscale or national treatment of the problem. The statistical work covering this region that consists of thirty-one counties with 20 million inhabitants is now nearing completion. It will be published in separate reports.[1] Thus, this paper does

[1] Forthcoming study by Joel Darmstadter of Resources for the Future, Inc.; Regional Plan Association, *Regional Energy Consumption* (Fall 1973).

not cite detailed statistics, but rather emphasizes overall relationships with the understanding that they are preliminary and tentative and do not reflect the views of the organizations involved in the study. Policy suggestions are strictly those of the author, and do not represent the findings of the joint project.

Levels of Consumption

Contrary to popular belief, per capita consumption of energy in the New York region is substantially *below* national averages, and per capita consumption in New York City is substantially lower than in the rest of the region. A resident of New York City consumes only half as much energy as his counterpart in the rest of the country, a resident of the suburban counties only two-thirds as much.[2]

Admittedly, much of this difference cannot be considered an inherent merit of the region. Having relatively little manufacturing industry, the region depends on imported products, for which energy is expended elsewhere. Energy-intensive users avoid the region because the cost, particularly of electricity and gas, is about 60 percent above national averages. Lastly, about half the difference in per capita consumption between New York City and its suburbs is a result of higher suburban income.

Nevertheless, part of the difference cannot be explained by any of these conventional reasons. It is due to an inherent merit of the region, namely, its high urban density. Consumption of energy decreases as urban density increases.

In the residential sector, the largest consumer of energy in the region, energy consumption per dollar of income declines noticeably with rising density. One might surmise that energy-use per square foot of housing area would also decline with higher density, because of smaller losses of heat to the atmosphere in apartments. That phenomenon, however, holds true only for a part of the density range—between the inner suburban counties and the city—and not for the outer counties.

In transportation, the second largest consumer of energy in the region, the trend is most sharply pronounced. No matter how it is mea-

---

[2] Resources for the Future, Inc. and Regional Plan Association, "Patterns of Energy Consumption in the Greater New York City Area, A Statistical Compendium," EN-2, mimeographed, July 1973.

sured—per capita, per dollar, or per square foot of floor space—energy consumption for transportation drops dramatically with rising density. People in a high-density environment travel less in general and more frequently use energy-saving public transportation.

Even in the commercial and institutional sector, which includes street lighting, differences by density are evident. Consumption per square foot of commercial floor space tends to decline as density rises. The same is true of industry, though in this case the trend reflects a tendency for energy-intensive users to locate farther from the center of the region.

Thus one can say without much exaggeration that high-density urban environments are energy-conserving environments. In guiding the future of urban growth, this consideration should be kept in mind. The dispersed development in the suburbs is more harmful to the environment than the concentrated development of the city.

While high urban density tends to reduce total energy consumption, it does increase the intensity of energy-release locally. For example, the release of man-made energy in Manhattan per square mile of land is about a thousand times greater than in the whole country and approximates the amount of energy the island receives from the sun. Yet because of natural ventilation, even this extreme condition causes relatively modest changes in microclimate; the weather is a few degrees warmer than in the surrounding area. This heat is not a total loss in a climate where the overall needs of winter heating far exceed air-conditioning needs. Of course, the situation is the reverse if one considers the electricity sector alone. Nevertheless, more harmful than the release of heat is the release of other energy by-products, notably noise and air pollution. In recent years it has been demonstrated that these annoyances can be cut back with proper technology and strict law enforcement. Even the Manhattan traffic could meet air-quality standards, if the proposed vehicle-emission standards are implemented in the 1976-86 period.[3] Thus the environmental cost of high-density urban areas seems to consist primarily of tighter waste-management standards and greater ingenuity in design, which seem worth the environmental savings of the reduced total consumption of energy. One might add that the high density of Manhattan has made possible a central supply of steam—a partial solution to the problem of waste heat in electricity-

---

[3] Jeffrey M. Zupan, *The Distribution of Air Quality in the New York Region* (Baltimore: Johns Hopkins Press, 1973).

generating stations—that is not feasible with a low-density settlement. Of course, energy use is but one of many social and environmental factors favoring higher urban density.

Rates of Growth

The growth rate of energy use in the New York region is also lower than in the nation as a whole. This fact, however, cannot be attributed to high urban density, for the density of the built-up portion of the region continues to decline. Rather, the reasons are primarily the population growth rate and the increases in per capita income, which are lower than the national average. Even though the pressures of a growing demand for energy are not as intense in the region as in the rest of the country, they are staggering in absolute terms.

In view of current trends, energy sources will be a greater concern for the next fifteen years than the increase of energy use. During the past decade coal, the most abundant domestic fuel, virtually went out of use in the region, even though it is still widely used nationally. It was replaced by cleaner but rarer energy sources, oil and natural gas. Many households and businesses shifted because of convenience. Electric power stations, however, shifted to these costly fuels to satisfy environmental standards. The introduction of nuclear fuels proceeded slower than expected; they produce 3 percent of the electricity, filling 0.25 percent of the total energy demand of the region.

Electricity production is expanding rapidly. Unlike the total energy growth rate, which is slower in the region than in the nation, the electricity growth rate is about the same. As in the rest of the nation, electricity use in the region, until most recently at least, has been doubling every decade. Relatively speaking, it is replacing other forms of energy use faster than in the rest of the nation, primarily because the region seems able to afford it. Although the region's income growth does lag behind the rest of the country's, the region's level of income is still about 25 percent above the average. Even doubling electricity generation every ten years could go on for a long time. Electric energy in the region today represents less than one-tenth of final demand. If, as is often suggested, the long-term trend is toward a predominantly electrical economy, no amount of conservation measures will resolve the difficulties of producing huge amounts of electricity in an environmentally safe way. At this point, a gratuitous thought might be added: while cooling water for nuclear plants is a constraint in many parts of the country, there is certainly plenty of it in the New York region, below the top thirty feet of the Atlantic Ocean. Water from this depth can be used without affecting the marine life that lives in the top stratum.

The use of gas in the region has not been expanding quite so rapidly as electricity. It also doubles every ten years. In the past two decades the old-fashioned gas works, which made gas out of coal, were demolished. The region began piping in superior natural gas from Texas, only to realize, rather belatedly, that its supply is not inexhaustible. It seems unlikely that the investment in the costly network of distribution pipelines will be abandoned in the foreseeable future. Nor is it likely that a clean fuel, which for a number of uses is more efficient than electricity, will be given up. Rather, along with conservation measures, new ways of mine-mouth coal gasification deserve serious consideration.

Liquid petroleum products, mostly the various fuel oils and gasoline, have a relatively low rate of growth, simply because the use of electricity and natural gas is expanding so much faster than energy use in general. But this sluggish growth rate should not obscure the fact that, in absolute numbers, by far the biggest increase in energy consumption in the region is in liquids derived from petroleum, which account for over three-quarters of the final energy use.

The increase is not caused by fuel oil, the dominant component of this energy form. In fact, the use of fuel oil is shrinking on a per capita basis and in some instances in absolute numbers as well. It seems to be repeating the cycle that coal went through a few decades earlier. The increase is caused primarily by gasoline for motor vehicles, which accounts for a huge share of the total energy demand in the region. To a considerable degree, it is also caused by kerosene for jet aircraft. The use of aviation fuel in the region more than tripled in the past decade and now accounts for about one-fifth of the energy demand for passenger transportation.

Energy for Passenger Transportation

A measure of travel demand closely related to resource use is passenger-miles of travel, which represents the total number of trips multiplied by their average length. This measure plays tricks on the unfamiliar reader, because it deflates the frequent, short trips and emphasizes the few that are very long. Bus travel, which may account for nearly one-tenth of all trips in the region, shrinks to a low percent by this accounting, because the average bus trip is only two to three miles long. Trips by air, which account for only a fraction of 1 percent of all trips, have great significance because the average air trip to or from New York is over nine hundred miles long. It is in this framework that table 1 presents the energy requirements of various modes of travel.

TABLE 1

Energy Use for Passenger Travel in the
New York Region in 1970

	Passenger-miles of travel		Share of energy used for travel (percent)	Calories consumed per passenger-mile
	(in billions)	(percent)		
Bus	3.38	2.5	0.9	560
Subway	8.75	6.5	2.5	635
Rail	5.70	4.3	1.7	645
Automobile	97.50	72.8	70.8	1,615
Airplane	17.66	13.2	22.6	2,770
Taxi	0.81	0.7	1.5	4,200
Total	133.80	100.0	100.0	1,665

Source: Regional Plan Association
Note: Data are preliminary and subject to minor adjustment.

It is evident that buses, subways, and railroads are indeed energy-conserving modes; they serve 13.3 percent of the travel demand in the region but consume only 5.1 percent of the raw-energy inputs into passenger transportation. The remaining 94.9 percent of the energy is consumed by automobiles, airplanes, and taxicabs, which furnish 86.7 percent of the passenger-miles.

However, it is the use of the energy-conserving modes that has been declining. From 1960 to 1970, bus travel in the New York region declined by about 14 percent (mostly outside New York City), subway travel dropped by 6 percent, and rail travel shrank similarly. The declines cannot be compared to the precipitous abandonment of public transit during the 1950s, but they do continue, despite intensified public capital investment.

Meanwhile, the use of energy-intensive modes, even taxicabs, keeps growing. It is most pronounced in air travel, where passenger-miles to and from the New York region more than tripled from 1960 to 1970, because of the time-saving nature of the jet airplane. More significant in absolute numbers, however, is the continuing expansion of automobile use; the number of motor vehicles in the New York region has been doubling every two decades since the Depression, and it continues to do so.

A decade ago the Regional Plan Association, in developing motor-vehicle projections, surmised that a saturation point in ownership must be near. Except in densely settled parts of New York City, this expectation has not materialized. On the contrary, in suburban counties of the region the increase in motor vehicles during the 1960s was greater than during the 1950s. More vehicles than people were added to the region in the past decade. Most of these vehicles were bought by families who already had at least one car. Thus, even as the number of private automobiles in the region during the past decade increased by 40 percent, the number of households without cars also increased, even if slightly. The growing inequality of income and the growing contrast between New York City and its environs is apparent in automobile ownership. One-quarter of the region's households now have two or three automobiles, while one-third remain without one, just as a decade ago.

With these trends continuing and the efficiency of the automobile going relentlessly down—a trend that has accelerated in the past few years—one can expect roughly a 70-percent increase in gasoline use in the region between 1970 and 1985. Moreover, with current growth rates, the amount of aviation fuel used would begin to approach that of gasoline for motor vehicles around 1985. By that time, the two transportation uses would account for well over half the region's energy consumption!

Clearly, such considerations shed a new light on the notion of building fourth and fifth jetports for freight as well as passengers. Questions of freight movement must remain outside the scope of this paper, but others have shown that throttling the growth of air freight would have a spectacular energy-saving effect.[4] Per ton-mile of freight, a jet airplane uses fifty-four times the energy of a railroad train.

Other steps, however, have to be taken closer to the ground. There is no shortage of suggestions. "Drive slower to save gasoline—fifty is thrifty" is one of them. Now, it is true that automobiles on the road

---

[4] Eric Hirst, "Energy-Intensiveness of Transportation," *Transportation Engineering Journal*, Proceedings of the American Society of Civil Engineers, 99 (February 1973), 111.

today minimize their fuel consumption between 30 and 50 miles per hour and that an increase in speed from 50 to 70 miles results in increased gasoline consumption of 45 to 20 percent.[5] But travel on main rural highways where speeds average 60 miles per hour accounts for only about one-third of all motor-vehicle travel in the United States. One-half of all travel occurs on urban streets, where speeds range between 22 and 26 miles per hour on the average and where a much greater fuel loss is caused by multiple stops and congestion. So, even in theory, savings from driving more slowly would be minimal. In practice, human behavior responds primarily to opportunities and needs, not to sermons. Numerous studies have shown how highly the traveller values saving time. To ask him to slow down when both care and highway design encourage high speed is unrealistic.

Car-pooling is another prescription suggested for some areas where public transportation is poor. However, car pools can only work when a group of people live and work in the same vicinity, such as a group of federal employees in Washington commuting from the same suburb to the same office building. In the New York region, such situations are fairly well covered by public transit. Cars are used only for trips where both the origins and the destinations are so dispersed that no other means of transportation is reasonable. Automobile commuters to Manhattan are relatively so few and live so far apart that, if pressed, it would be to their advantage to use public transit before considering car pools. Lastly, there is also the psychological problem of urging car pools upon a population that is buying second and third cars.

Physical exclusion of automobiles from selected areas, such as parks or downtown districts, which are accessible by alternate means of transport, does offer some possibilities. The theory is that traffic rises to fill the available pavement. Thus, added pavement produces "induced" traffic, which did not exist before and, conversely, less pavement will not just divert traffic to parallel facilities, but will suppress some traffic altogether. There is empirical evidence to verify this theory, the application of which can be environmentally as well as psychologically desirable. However, such measures will not save much energy because there are few areas where streets can be closed in the region. For example, banning all cars and taxis from the central square mile of mid-Manhattan would reduce total vehicular travel in the New York region by less than 1 percent. The defeat of the more modest

[5] Paul J. Claffey, *Running Costs of Motor Vehicles as Affected by Road Design and Traffic*, National Cooperative Highway Research Report 111, Highway Research Board, 1971.

Madison Mall proposal by the New York City Board of Estimate demonstrates the political difficulties of such measures.

At the same time, broader and societywide solutions are advocated. "Tax the hell out of the automobile" is one of them. Yet, once again, reality is more complex than slogans. First, the taxes would have to be very, very high to affect the behavior of those persons buying second and third cars, and one would have to be careful to see that they do not deduct them on their income tax. In the days when public transit was dominant and the automobile was an auxiliary travel mode, a round trip on the subway cost 10 cents, and a round trip across the George Washington Bridge cost $1. Today, the latter still costs $1, while the former costs 70 cents and may be $1 before long. Demand-elasticity studies suggest that the difference in price would again have to be about tenfold—a $7 to $10 toll on Hudson River crossings—to significantly cut back demand. In view of the political resistance to even minor toll increases, it is doubtful that elected officials will soon take such measures.

Second, and more important, extreme measures in this direction would effectively preclude the use of automobiles by people with low incomes. Given the continuing dispersal of blue-collar jobs, the car is the only way for wage earners of many low-income households to get to work, and it is their major opportunity to break out of the ghetto, as a recent Regional Plan study demonstrated.[6] Thus a society that wants to use the pricing mechanism for energy conservation must either achieve a much greater degree of income equality, or it must devise a selective system of pricing and taxation that is not regressive.

Since the tendency, though not the rhetoric, of the 1960s and 1970s has been toward greater inequality, highly selective measures must be pursued in the short run. The most promising avenue is a steeply progressive tax on the horsepower or the size of cars. The 1,615 calories per passenger-mile indicated in table 1 as the energy consumption of passenger cars in the New York region reflect an average efficiency of 13 miles a gallon with 1.5 passengers per car and the existing predominance of large cars. Small cars on the road today have no trouble achieving twice this efficiency. As is evident from the table, this efficiency, with around 800 calories per passenger-mile, comes close to the efficiency of electric rail transit, which is around 640 calories a pas-

[6] Regional Plan Association, *Transportation and Economic Opportunity: A Report to the Transportation Administration of the City of New York*, no. 119 (New York, 1973).

senger-mile, because transit wastes two-thirds of the fuel in the process of converting it into electricity.

If twenty-one miles a gallon was established as an efficiency goal for the average car in the region and phased in over the next twelve years, by 1985 there would be no increase in gasoline consumption even with the projected increase of 3 million more automobiles. In the process, there would be no disruption of established patterns of travel, and a modest degree of income redistribution would be achieved. However, such measures would sacrifice the large car—perhaps the prime symbol of the American dream.

The author does not wish to appear skeptical about public transit. It is energy-efficient and has other social and environmental advantages. The Regional Plan Association has been in the forefront of the effort to improve public transportation for over two decades. But the limitations of a shift to public transit as an energy-saving device must be realized. During the same two decades the region committed itself to a spread and dispersed pattern of settlement, most of which can be served only by individual vehicles. If operating subsidies are enacted to prevent further increases in transit fares, the region may be able to avoid further declines in transit use. To expect large increases in riders in the near future is unrealistic because of the committed pattern of land-use and because new transit capital-construction projects, except in the bus area, take about twenty years from plan to reality. Added subsidy and reduced fares can attract new transit riders, as Atlanta has shown. Calculations show, however, that the attracted riders are primarily new ones who did not travel before, not motorists lured away from their cars. To divert the motorist, barriers are needed, such as street closings, tolls, and parking surcharges. These, however, will affect only the small part of automobile traffic that travels in territories where the automobile and public transit compete. Thus, in the near future, diversion to public transit will remain only a modest component of a fuel-conserving strategy.

In the long term, of course, one hopes that it will be realized that high-density urban environments are energy-conserving environments: high density automatically encourages public transit and discourages the automobile. Whether the region and the nation will have a chance to rechannel urban growth into more compact urban forms remains uncertain. If the current decline in fertility rates continues unabated, zero population growth will begin in the New York region by 1985, and the present urban form will remain.

The Need for
a New Perspective on Energy

FRANCIS W. SARGENT

America faces a crisis brought about by a life-style born and nurtured in an atmosphere of abundance. Americans have always had an abundance of everything—land, minerals, food, and energy. It is a humbling experience for America to admit that its resources are limited. Yet the nation's future is dependent on its ability to deal with this reality.

Nowhere is this more true than in the area of energy resources. Certainly, it is conceivable that the projected energy demand could be met from now until the year 2000 by simply developing the necessary supplies. But, in dollars and in effect on the environment and the very quality of life, the price would be prohibitive. To meet the minimum demands projected for electricity in the year 2000, for example, a new power plant would have to be brought on line every three weeks from now until the year 1985. And further, from 1985 to the year 2000, the pace would have to quicken to a new power plant each week.

Such consequences inevitably lead to the conclusion that the energy needs of this nation cannot always be met by building another oil well or another power plant. For eventually there will come a day when the price tag becomes too great.

That day is fast approaching. Consequently, the nation must begin to develop rational energy policies which recognize that the present energy shortage is not so much a problem of supply as one of demand.

I do not in any way advocate that we ignore the need to generate an additional supply—for such a policy would be suicidal. However, it is necessary to understand that the ultimate solution to the energy problem lies, to a large extent, in an ability to manage the rate of consumption.

Details of the demand rate are well known. While the United States possesses only 6 percent of the world's population, it alone accounts for 30 percent of the total fuel consumed on this planet. This nation's rate of consumption is now expanding three times faster than the population. America is like an addict caught in an endless cycle: the more energy it gets, the more it demands.

To some extent, this growth in demand will be curbed by substantial increases in the price of energy products. Recent negotiations with the producing nations seem to indicate that further increases in the cost of crude oil will be forthcoming. With the potential for price hikes as great as it is, the law of supply and demand will not be denied: demand for oil will be moderated.

However, the nation cannot wait passively for the law of supply and demand to take effect. To do so would invite major economic dislocations. We must take the initiative to reduce the excesses that are so common to this nation's use of energy. To do this we must focus our attention on two of the most energy-inefficient creations in the history of man, the automobile and the modern house.

The United States presently consumes 55 percent of the world's gasoline. Between 1960 and 1970, consumption of gasoline as a highway motor fuel increased 64 percent, a rate of consumption expected to jump another 85 percent by 1985. These figures carry even greater weight when one considers that for every five gallons of fuel we put into automobiles, four are wasted. In fact, one resource expert has stated that only 5 percent of the gasoline consumed by automobiles is actually used in moving passengers.

If Americans are ever to control this nation's runaway demand for energy, they must, therefore, attack this waste at its source. Reliance on the automobile and the highway must be replaced by more energy-efficient modes, such as railway and bus.

As governor, I have begun this task in Massachusetts. I have said that the state will no longer allow multilaned expressways to smash into Boston. Instead I have proposed a $2-billion transit program designed to give people an effective alternative to automobiles.

I have introduced legislation to have the state assume, out of general revenues, half the cost of the deficit incurred by the transit system. This action will help hold down fares without putting an added strain on the budgets of local cities and towns.

I have also introduced and supported legislation authorizing the creation of regional transit authorities in other areas of the state; for it is clear that Boston is not the only city in Massachusetts in need of mass transit.

Finally, for the last three years I lobbied hard in support of recently enacted legislation that has revised and liberalized the national Highway Trust Fund. In many ways, this legislation can be considered as the most important energy proposal Congress has dealt with this year.

To be sure, there are some who argue that public transit can never be a real alternative to the automobile, that the average American is too attached to his car to give it up. Yet, there are some statistics that seem to say that this "love affair" is more like a chilly marriage of convenience.

Three American cities, San Diego, Atlanta, and Denver have lowered their transit fares in an attempt to increase ridership and reduce the use of the private automobile. The results exceeded all expectations.

In San Diego, the local transit authority abandoned a complicated rate structure based on zones in favor of a single fare. While the rates under the old system varied from forty to ninety cents per ride, the new system established a single trip fare of twenty-five cents. The result: a 72 percent increase in patronage.

In Atlanta, when fares dropped from forty cents per ride to fifteen cents, patronage jumped 23 percent, a figure that has been maintained for two years.

Boston has just initiated a program called "Dime Time" which allows passengers to ride the transit system during certain hours of the day for ten cents instead of the usual fare of twenty-five cents. This experiment has greatly increased the number of riders during those hours.

Boston is also beginning to experiment with monthly tickets. In return for a voluntary monthly payroll deduction, workers will receive an unlimited monthly pass to use the transit system.

When one considers these experiments—especially in light of the energy crisis—it seems evident that government, both federal and state, must give serious consideration to no-fare public transportation. Certainly, this is a controversial proposal, but prepaid transit may well become a reality in the foreseeable future.

Even if an effective system of mass transit can be developed throughout this country, many Americans will still be dependent on their automobile. An individual cannot be deprived of his car if he has no other mode of transportation. The question then is what kind of car is America going to use. Is it going to be a 5,000-pound, air-conditioned luxury car? Or will Americans move toward more energy-efficient automobiles? Unfortunately, the trend to date has not been in that direction. In fact, for 1973 American cars, the average fuel consumption is about thirteen miles per gallon. There has even been one incident of an American luxury car averaging three and one-half miles per gallon.

One way to reverse this trend would be to provide a sliding excise or sales tax, whereby the owners of more energy-efficient cars would pay less than the owners of energy-inefficient cars. The key would be to make the tax high enough to persuade a prospective purchaser to think twice before buying a luxury car. Revenue derived from this tax could be used for further research into making automobiles less polluting and more energy efficient.

Now I have said a great deal about transportation, but energy waste is not limited to automobiles. Reducing reliance upon the automobile will, of course, result in reducing energy consumption. But to do the entire job, we must also examine our homes, which are blatant examples of how to waste energy.

Americans have the naive notion that the cost of a house, a car, or an appliance stops on the day of purchase. This notion was nurtured in an era of abundant energy supplies and low energy prices. Abundance, however, is not the case today. The reality is that quite frequently the cost of operation easily exceeds the price of purchase.

One way to control such excesses is to inform the consumer of the hidden energy costs. For once the American consumer realizes what he is really paying, he will begin to insist on energy-efficient products.

To begin to provide this information to the consumer, I submitted legislation in Massachusetts to require labels stating the operating cost of certain major appliances. Recently I signed this legislation into law, making Massachusetts the first state in the nation to adopt such a program. However, we must go beyond this law. Programs must be developed to guarantee the availability of such information for cars, for houses, and even for office buildings.

If one knew that he could cut his heating bill 42 percent a year by insisting on minimal insulation, such insulation would be installed. Presently there is no way the average citizen can be informed of these facts unless he specifically requests this information. As with appliances, it will require the action of government to ensure that each consumer understands the full dimensions of his purchases with respect to energy.

President Nixon has recently established a Bureau of Energy Conservation in the Department of the Interior. The obvious purpose of this bureau is to develop programs to conserve energy. But it should do more. It should be integrally involved in all energy planning. The nation must not simply be concerned about how to develop the supplies of energy to meet demand, but how to limit demand to meet supply. This requires some consideration of an equally critical management problem—that of supply.

In discussing this problem, one must be cognizant of two essential factors. First, there is no worldwide shortage of energy supplies. The truth is that there are ample petroleum reserves throughout the world to meet demand at least through this decade. The immediate problem is one of obtaining and distributing these supplies. This is the challenge now confronting the major consuming nations.

The immediate energy shortfall did not suddenly develop in an unsuspecting oil industry. Their economists could have easily foreseen recent developments. It is logical then to ask why these companies did not act to meet the increasing demand.

The most likely answer is quite simple. For the industry to have met these needs, it would have had to increase substantially its costs and subsequently its prices when Americans were not prepared for these events. The result would have been lower sales, less profits, and a lot of angry consumers. It should be pointed out that wage-price controls only exacerbated this problem by further constraining the flexibility of the oil industry.

In truth, the companies, with the help of the federal government, seem to have "managed" themselves and the American public into a very tight situation. A good example of this attitude is that before the summer of 1973 the major oil companies claimed it was impossible to work refineries at anything more than 90 percent of capacity. Today, with higher prices and higher profit margins, they are refining at close to 96 percent capacity.

It is clear, therefore, that one cannot and should not expect the energy companies to be the guardian of the public interest. Nor do I mean to imply that oil and gas companies are out to exploit the consumer. They are simply profit-maximizing institutions operating in a quasi-monopolistic market, and they are going to act as such.

If the companies are not going to protect the American consumer, who is to assume this responsibility? I believe it must be government—both federal and state. In fact, I believe that we will soon see the day when the major oil companies will join their electrical colleagues under the jurisdiction of strong government regulation.

In fact, upon close examination, it becomes quite evident that the oil industry is already regulated by the federal government. One has only to look at the multitude of federal permits, licenses, and tax programs governing these companies to be convinced that the industry is now regulated. The only difficulty is that, as one would expect, the federal government does not admit to this regulation; for as soon as it did, the general public would direct their dissatisfaction at Washington, not at the industry—a situation any administration would prefer to avoid.

Already the ramifications of this position have become evident. The federal government, for example, delayed implementation of the Mandatory Allocation System not because it had any philosophical objections but because it did not want to place itself in a position where it would be held responsible for the maintenance of fuel supplies.

As a result, federal energy policy up to this year has been crisis-oriented, totally disjointed, without continuity. Instead of a consumer-oriented approach, national policy tended to represent the short-term interests of the oil industry, with little regard for long-range consequences. Additionally, programs and policies once instituted were seldom reevaluated, often becoming destructive as factors changed. The oil-import quota system, for instance may have had merit in 1959, but by 1968 it was obvious that changes were required. Despite this need, five years passed before these changes were made.

Federal mismanagement of energy policies has proceeded through several administrations, beginning with the Eisenhower administration, which basically let the oil companies dictate federal policy, through the Kennedy-Johnson administrations, which managed to lay the groundwork for the present natural-gas problems, to the Nixon administration, which until this year has been reluctant to confront the energy issue.

This federal mismanagement has victimized not only the American consumer, but the industry itself, whose range of action has been limited by the instability of federal policy. Private companies have consistently avoided substantial, long-term capital investments when the real possibility has existed that new federal policies would make such investments unproductive. As long as there was doubt as to the future of the oil-import quota, for instance, the oil industry was reluctant to commit itself to new refineries.

When the nation enters the second part of this decade, the need for continuity in federal policy will become even more critical. During this period one can reasonably expect that future costs of petroleum production, except perhaps in Kuwait and Saudi Arabia, will increase rapidly, if for no other reason than the decreasing accessibility of new supplies. One can also expect that the willingness of the oil industry to make capital investments to obtain fossil fuels located in increasingly remote areas, such as the outer continental shelf, will rest to a large extent on how stable the industry perceives federal policy to be.

The long-term task, therefore, must be to structure a federal regulatory role that can balance the many complex and demanding factors involved in this field. All of us must work to ensure that the government has the capability and the willingness to act to protect the interests of the general public.

Until this year, the federal government lacked both qualities, and only recently did the president make it clear that the government intended to take a strong role in energy matters. Unfortunately, however, the atmosphere of crisis that has descended upon this issue will now make it most difficult to build up a capability to handle energy issues in an objective and effective manner. As I write these words, this nation stands on the verge of a heating-oil shortage of grave proportions, a situation compounded by the crisis in the Middle East. As one could expect, the attention of federal and state officials is centered on one immediate issue alone: surviving this winter. As the inevitable complement of this, long-range problems such as the orderly establishment of regulatory machinery are being lost from sight.

Despite the very real pressures under which public officials now find themselves, certain basic actions can be taken that will begin to shape a new role for the federal government in the field of energy. First and foremost, the federal government must demand accurate and comprehensive information from the oil corporations. Last year when Massachusetts was informed by major suppliers of heating and residual oil that they were in short supply, no one in Washington knew or could even estimate whether the company statements were accurate, or whether additional stocks of fuel were anywhere to be found. In fact, my staff found that brokers on Wall Street had more information on what was going on in the oil industry than officials in Washington.

I was so appalled by this situation that I initiated legislation in Massachusetts to require oil companies doing business in the state to submit pertinent information on supplies. On October 26, 1973, I signed this legislation into law. I believe Massachusetts is the first state to have taken such a step. Meanwhile, under the leadership of the president's adviser, John A. Love, the federal government finally began to demand more information from the oil industry.

A second action that the federal government should take is to see that the information received is subjected to closer scrutiny. Since presently most of the data employed in decision-making comes from the industry itself, the major issues affecting the energy industry are decided on information supplied by the industry. I know, as a governor, that it is extremely difficult to determine policy based on data that come from vested interests in the decision. It is vital, therefore, that a reliable and credible base of information be developed on each energy issue through means independent of the involved corporations. Only through such independence can the regulators be sure that the information that industry does provide is valid and that these data can safely be used in decision-making.

Third, the federal government should commit itself to the reevaluation of all existing and proposed energy programs in light of this increased information base. The world situation is changing so rapidly that policies effective one year can be ineffective the next.

Fourth, the nation must accelerate its research and development programs. The president has taken vigorous action to provide for such an effort by proposing that the federal government appropriate $10 billion over the next five years for energy research. I support this commitment and urge that these moneys be channeled into solar and fusion research as well as the more conventional alternatives such as coal gasification and breeder reactors.

There is, however, a pressing need to go beyond this long-term commitment. Short-term problems demand immediate action. This winter, for example, under the Mandatory Allocation Program, each governor will have a supply of heating fuel available with which he must attempt to meet the energy demands of his state.

In carrying out this program the governors hopefully will work to develop an energy-conservation program that is as effective as possible. In doing so, they will confront an imposing range of questions that have no answers today. What actions conserve the maximum amount of energy with the minimum amount of social disruption? If rationing becomes necessary, how does one minimize inefficiencies? One way to discover the answers is by trial and error; another is through advance research. Clearly this latter approach is more advisable and should be adopted immediately.

Finally, the federal government and the states must commit themselves to make decisions on energy matters in an open and participatory manner. Closed-door methods only breed distrust, and this nation can no longer afford such reactions. Consequently, every effort must be made to ensure that state, regional, and local entities, as well as private organizations, have the opportunity to participate in the decision-making process.

I realize that some of these suggestions may be unpalatable. But the nation is now faced with a potential crisis, the dimensions of which cannot be minimized.

America is the richest and most powerful nation in the history of man, a nation that has never known the word *limit*. With boundless confidence that we can overcome any obstacle that may be thrown our way, we become very frustrated if we are unable to find quick solutions to our problems.

It is not easy to convince people who have never been without something that more of the same is not always the answer. Nor is it easy to tell this nation that part of its economy must be regulated and managed by the government, for we believe that the free-market system is representative of all that is good in our economy while government intervention remains at best a distasteful necessity. Yet if the nation is to survive the energy crisis, it must face these new realities; for the price of continuing on the present course is much too high.

Thus it is up to government, at both the state and federal levels, to seize the responsibility of leadership in developing objective and effective energy policies—policies that will be directed at managing both a diminishing supply of resources and a spiraling demand.

Unless Americans realize that the energy problem is a two-part problem, one of supply and one of demand, they will find themselves confronted with ever-increasing shortages. Unless they acknowledge the dual nature of the energy equation, they will one day awaken to find themselves permanently disabled by their own shortsightedness.

Energy and Society

MARVIN J. CETRON
VARY T. COATES

One forecast of life at the end of the twentieth century can be represented by the Alpha family scenario. A conspicuous aspect of this life-style is an increased consumption of energy by the use of electronic cleaning devices and disposal and security systems.

The Alpha home is a unit in a condominium constructed in the early 1990s of prefabricated modules assembled on site. It is built in the style made famous by Habitat as early as 1967 but not generally accepted at the time, in part as a result of security problems that have since been solved by the use of electronic monitoring of all entrances and passageways.

The Alphas' home is heated and cooled by an electric system, and the air is electronically cleaned. Cleaning chores are further reduced by a central vacuuming system with additional cycles for wet cleaning, such as scrubbing floors and shampooing carpets. The house is, of course, equipped with a disposal appliance, which receives all household trash and garbage, sorts it with chemical sensors, crushes and compacts recyclable materials into separate containers to be picked up by the city's sanitary-engineering service, and reduces the remainder to biodegradable powder which is flushed away with other organic wastes to be fed into the agricultural station nearby.

A different forecast postulates an energy shortage. Gas is rationed. The use of electricity is curtailed. City housing is overcrowded.

Trips to shopping centers, once considered nearby, must be carefully planned and rationed because of the cost of gas. Home air conditioning is not much used. Bills are too high to use electricity routinely, and on very hot days there are often brownouts and blackouts. Many houses in distant suburbia have been abandoned because of the chronic shortages and high prices of gasoline.

In both scenarios it is evident that the supply of energy has an effect on how people live. But the scenarios also reflect the inevitable impact of social attitudes on the uses of technology. Underlying the first scenario is the assumption not only that the future supply of energy will be adequate, at least for the affluent Alphas, but also that such conspicuous consumption is both accepted and encouraged by society. The second scenario reflects not only a shortage of energy, but also a different attitude toward consumption.

The interplay between attitudes and technology occurs at one stage in a three-part process of technological change. The initial phase of innovation involves technical feasibility. It occurs when, with exacting specifications, one material or process can be substituted for another. For example, the use of plastics as a substitute for metals is technologically feasible in the automobile industry. In fact, by 1984, 50 percent of an automobile by weight will be made out of plastic, not steel. By 1990, 92 percent will be made out of plastic. The technical requirements for automobile construction indicate that plastics can be substituted for steel.

The second stage involves economic feasibility. No competitive business could change its construction materials if it was not cost-effective.

The third stage of change, which involves social attitudes toward technology, is the key to the whole process. Ironically, social attitudes are sometimes overlooked when introducing change. Five years ago, Sweden created a special road for buses in Stockholm. The government took a good three-lane highway and reserved two lanes for automobiles and the other lane for buses. Using its special lane, the bus can take Swedes downtown in 20 percent of the time it takes by automobile. Although the bus is free, few people take it. The Swedish government had not sufficiently taken into account commuters' attitudes on schedules, subsidized transport, and freedom of movement. For these and probably other reasons, the Swedes did not consider the bus experiment to be an acceptable innovation.

One reason why social attitudes are overlooked is that they are more difficult to predict than technological and economic feasibility. It is possible to chart the course of substitution for plastics and other materials in the automobile industry very accurately. Using substitution curves, forecasters can mathematically relate a material to its substitute over time.

Although social forecasting is much more difficult and experimental, it is possible to make social predictions by using some of the concepts of technological and economic forecasting. One useful concept is the notion of "precursors." In technological forecasting, precursor events can be used to predict change. If it is known, for example, that x precedes y, then it is possible, given the occurrence of x, to predict the coming of y. Similar methods are applied in economics. For example, the United States trade deficit was seen as a precursor of the devaluation of the dollar. In social forecasting, predictions can be made using precursor nations. Scandinavian countries, for instance, have been precursors of many changes in the United States. Free medical care was instituted in Sweden years before it was in the United States. Likewise, young unmarried Scandinavians started living together years before their American counterparts adopted the same custom.

Using Scandinavia as a precursor reveals several social innovations as probable future changes in the United States. Nationalization of telecommunications and transportation industries, an established Scandinavian practice, has just begun in the United States with the coming of Amtrak. The practice can be expected to continue in the transportation industry and to spread to telecommunications facilities as well. No-fault insurance, already instituted in Sweden, is now the subject of political debate in the United States Congress. It is probably only a matter of time until it will be the legally accepted norm. Likewise, no-fault divorce, a logical cousin to no-fault insurance and an existing Swedish practice, will probably arrive in the United States within a decade.

Once it is possible to make reasonable statements about future social change, forecasters can combine these predictions with technological and economic projections to produce future scenarios. The value of scenarios is that they describe future environments under varying basic assumptions. The following scenarios depict two possible ways that the supply of energy may affect society. The Alpha scenario assumes a high level of energy consumption, combines this with predicted changes, and describes the resulting environment.

Communications have undergone several changes, including increased consumption of energy. The most visible and most used appliances in the Alpha household are, of course, the various telecommunications devices, such as the interactive two-way cable television con-

sole and the videophone, both of which can be plugged into the facsimile machine. Mail and newspapers are scanned on the television screen and, when a copy is desired, printed out by the facsimile machine. Most household shopping, except that considered as recreation or socializing, is also done by two-way television. The Alpha children use the resources of the regional information center (still often called "the library") through these communication devices.

Housing arrangements, made possible by innovations in transportation and construction, also reflect the impact of new energy sources. The Alphas' second home, in a remote area, is used several days of the week. The family commutes between their two residences by STOL plane from a nearby airport. Small commuter buses run between the airport and home. The second home, a mobile unit, is currently located in an area that was opened to development during the previous decade after a large desalination nuclear-power plant was constructed. The area, once an uninhabitable desert, now combines low-density residential clusters with controlled-environment farming and fish-culture, yeast-culture, and food-processing plants. The agricultural complexes are serviced by automated irrigation systems.

Work patterns have also changed. The active work force has been greatly reduced as a result of continuing automation, so that blue-collar workers and the lower ranks of white-collar and service employees usually retire or shift to part-time specialized occupations by their mid-fifties, encouraged by social-security incentives designed to distribute available work.

The Beta family lives in a very different world characterized by an acute shortage of energy. The shortage has widespread repercussions. Heating houses is a constant problem.

The Beta house is equipped with an oil furnace, which is a source of difficulty since heating oil is now rationed. The Betas have invested heavily in insulation and storm windows. By keeping the thermostat set somewhat lower than the elderly members of the family find comfortable, it is not difficult to keep the house tolerably warm all winter, except for occasional emergencies when the oil company runs short of supplies or its trucks are out of commission for lack of gasoline. Unfortunately, the Betas cannot adapt their town house to residential solar heating as many of their friends have done, since neighboring apartment houses block off direct sunlight for too much of the day.

Transportation also suffers from the rationing of gasoline. For air travel, trips need to be scheduled weeks, sometimes months, in ad-

vance. Excess competitive flights have been reduced so that scheduling is less convenient than in the past, and planes fly with full passenger loads on most flights. The expected boom in air freight during the 1980s did not materialize, since airplanes are inefficient users of energy and costs are therefore high.

Productivity in the Beta world also reflects the impact of the energy supply. Mr. Beta's work problems are magnified because labor is undependable. Workers run short of gasoline, and their cars are stranded or delayed. Shortages of energy slow production intermittently, and inevitable arguments between managers and government regulators over material and energy priorities further interfere with efficient production.

The relationship between life-styles and energy demand in the United States presents a curious problem. The low cost of energy in the context of an affluent society and its general availability mean that energy has been a minor consideration in determining life-style, even for the poor. Because the use of energy permeates every facet of American life, any constraints imposed by energy shortages—such as high prices, conservation measures, and rationing—are apt to bring about disproportionate consequences in the form of resentment, political pressures, conflict of interest, polarization, and economic dislocations.

Many other aspects of society can be similarly described in scenarios, including the international ramifications, such as the impact on the balance of trade and payments and changes in diplomatic relations with other nations. This process of considering the many-faceted consequences of technological innovation for society is called technology assessment. Such an assessment seeks to identify both beneficial and harmful impacts of technological change on societal structures, institutions, programs, and people. Scenarios, though an important part of technology assessments, are only the beginning.

In an onging project for the National Science Foundation, Forecasting International and General Electric Tempo are investigating the societal impacts of various solar-energy systems. The first phase of the project focuses on the development of scenarios similar to the Alpha and Beta examples, though with an emphasis on the impact of specific systems. Into these scenarios will go detailed forecasts of technological advances in solar energy and predictions of likely economic and social changes. When the scenarios are completed, they will form the basis for the actual impact assessment, phase two of the project.

In order to systematically assess the impacts of various solar-energy systems, Forecasting International and General Electric Tempo will conduct a survey of the opinions of experts in several fields. Some of the fields to be represented include economics, sociology, environmental law, environmental sciences, agriculture and land-use, public administration, government-industry relations, science policy, international law, architecture, and community planning. The framework for their responses will be a structured questionnaire. When the data from all of the questionnaires have been tabulated, the project team will construct a summary matrix for each solar-energy system as shown in table 1.

In addition to identifying all affected parties, the goal of the survey is to collect information that will indicate the probability of the impacts occurring (under varying assumptions); the direction of change they cause (increase or decrease); the magnitude of that change (percentage, proportion); and the duration of the change. Since the art of technology assessment is still highly experimental, the impact information may be fragmentary. Yet it is an important beginning.

TABLE 1

Sample of Groups Affected by Changes in Given Energy-Use Systems

| *Application* | Impacted Groups | |
	Direct	*Indirect*
Space heating and cooling	Home buyers	Mortgage banks
	Home builders	Trade unions
	Utilities	Building suppliers
	System manufacturers and distributors	Fuel suppliers
Electricity generation	Utilities	Fuel suppliers
	Environmental interest groups	Congress
	Users	Taxing jurisdictions
	Industry	Regulatory agencies
	Appliance suppliers	
Automobiles	Automobile manufacturers	Fuel suppliers
	Automobile users	Automobile service industry
	Environmental interest groups	
	Urban residents	
Mass transportation	Commuters	Utilities
	Nondrivers	Urban planners
	Local government jurisdictions	Real estate companies
	Taxpayers	Employers and employees

Government officials have begun to realize the importance of considering the kind of social information that a technology assessment generates. In addition to its solar-energy project, the National Science Foundation is sponsoring several other energy-related assessments, including one on off-shore oil, one on geothermal energy, and one on energy conservation. Congress, undoubtedly awakened by the nature of the public debate over the Alaska pipeline, has concurrently established the Office of Technology Assessment with a mandate to perform studies on the societal impacts of various innovations that are the subject of congressional inquiry.

It is not by chance that the government is interested in societal impacts. Government concerns appear to go through a three-stage process. At the most elementary level, the state, like the individual, must ensure survival. People are first concerned with making enough money to buy adequate food, clothing, and shelter. Likewise, governments seek to provide adequate housing, provisions, and employment. At the second stage, concern shifts to long-term growth and continuity. Individuals begin to seek better housing and better jobs, at the same time protecting what they have already acquired. Governments begin to invest in research and development, as well as training programs, and to emphasize protective measures, such as the development of elaborate military systems. In the United States, this second phase began sometime after World War II when the American economy began to experience unprecedented growth. In the past decade, the United States entered the third phase and initiated societal programs. Day-care centers opened. At least 4 to 5 percent of employees of corporations and agencies were members of minority groups, and about 1 percent was disabled. As pollution became an issue, companies began to feel pressure to cease dumping effluents into rivers, lakes, and streams.

At the time these social concerns were becoming legitimate considerations for policy-makers, the United States began to experience its first major resource shortage. Energy, previously viewed as infinite, began to seem a very finite commodity. Americans experienced their first peacetime shortages of gasoline. Social values, such as environmental quality, clashed with the need for energy, which is usually considered a necessity for the survival of American life-styles. United States policy-makers had to balance concerns for energy and the environment. In some cases, energy needs won out with only lip service to the environment. The oil shortage in Maryland, for example, caused the government to increase the amount of sulphur that can be emitted into the air when coal is burned. The Alaska pipeline, once rejected for environmental reasons, was approved by Congress with little public outcry.

Such policies will remain the order of the day in the near future. To make decisions intelligently, officials must consider all aspects of the problem. In a democratic system, the impact of decisions on various segments of the population must be considered along with the requirements for added energy. Who gets hurt must be considered along with what is needed economically. Ideally, technology assessments can provide policy-relevant information, which identifies where impacts are most severe and suggests alternative programs that alleviate these undesirable effects. This is the objective of the NSF project on solar energy, as well as of the research to be conducted by the Office of Technology Assessment.

The need for this kind of information can be readily seen. One current, classified contingency plan for dealing with the energy crisis suggests a return to gunboat diplomacy. Made public by an irate senator, William Fulbright, the plan called for United States arms shipments to Iran for the ostensible reason of encouraging the shah to extend his political control to include the oil-rich lands in Saudi Arabia and Kuwait. The plan even envisioned a deal to proportion the captured oil among the USSR, Europe, and the United States.

In fact, the plan has a certain appeal, and it has not been completely ruled out by all observers. However, it would be foolhardy in the extreme to implement it, morality aside, without first considering its impact on attitudes in the United States. For those who are concerned about the effect that Vietnam has had on American society, the plan must represent a chilling scenario for meeting energy needs. And who knows how the shah would act if he controlled the Persian Gulf oil?

Any way that the problem of energy is approached, from diplomatic-military contingency plans to solar-energy systems, it has an undeniable relationship with the social environment. What is required to deal with these needs is not only innovation, but also an understanding of the interplay between technological change and society. Consideration of social impacts and attitudes, for which technology assessment is one possible technique, coupled with imaginative and more effective applications of technologies can provide ways of dealing with energy needs that avoid the repercussions of international dirty tricks. In the future, as more is learned about technology and society, governments should also find it possible to manage other resource and environmental problems without inflicting major damage on the people in whose name it seeks these solutions.

Shortage Amid Plenty

WILLIAM D. SMITH

According to an adage among policemen and news-
papermen, there is no such thing as a free lunch. It is an old saw but
one that is still sharp, and it applies quite accurately to the nation's
present energy situation.

Americans, since the days of the Pilgrims, have enjoyed a virtually
free lunch with regard to energy. They have squandered fuels with
a profligacy that might well have brought down the wrath of the
Old Testament God. The United States grew up and became a super-
state on a diet of cheap, plentiful indigenous energy resources. With
6 percent of the world's population, it accounts for almost 33 percent
of the globe's energy consumption. This ready supply of energy has
played a major role in making America the world's largest industrial
complex and in giving its citizens, even its poorer ones, a standard
of creature comforts unrivaled in history.

From New York to San Francisco, the average American is a prodigious user of natural resources. With regard to energy, he floods his
home with light even when no one is in, heats rooms until they are
oven hot, and drives a gas-devouring car for a pack of cigarettes rather than walk a block. In Texas women carry full-length mink coats in
August to watch a baseball game in a huge air-conditioned dome.
And then there are electric toothbrushes, combs, tie racks, and hair
dryers.

A vast storehouse of coal, oil, and gas, which catered to creature
comforts, also allowed the United States an unequalled independence
in foriegn policy. Of all the major industrial countries in the free

world, America alone was self-sufficient in energy. But it is no longer in that position, as the Arab oil embargo has clearly demonstrated. Now only China and the Soviet Union among the major world powers are totally independent of foreign control over energy needs.

America is now beginning to pay for its lunch. Payment has come due earlier than expected because of the Middle East war, but the bill would have been presented soon anyway. There were small dislocations in gasoline supply during the summer of 1973, bringing muted moans from the public and loud howls from politicians. The Richter Scale of political complaint will reach a new high as shortages of heating oil, gasoline, and other fuels occur in the winter of 1973. If the statistics concerning the supply and demand of available energy resources are to be believed, things are going to get worse before they get better. Of course, statistics are not always to be believed, but even when they should be believed people do not always believe them.

Mankind has used more energy in the past thirty years than in all its history prior to 1940. The apex of this energy orgy is yet to be reached, according to many experts. World energy consumption is expected to double between 1970 and 1980, increasing from the equivalent of 87 million barrels of crude oil a day to around 150 million barrels a day, as the population grows and less industrialized nations strive to develop. Energy consumption in the United States doubled between 1950 and 1970 and is expected to double again between 1970 and 1985, increasing from the equivalent of about 31.8 million barrels of crude oil a day to 62 million barrels a day.

Most of the figures available are from industrial sources, and their perspective can therefore be considered one-sided. The trend, however, based simply on population growth and the aspirations of the have-nots, appears inescapable. Rigid conservation programs could reduce growth, but they cannot turn back the clock.

Yet there is no physical shortage of energy resources in either the United States or the world. The United Sates has the basic energy materials to meet its needs for at least 200 years, according to the National Petroleum Council. The council, which is an industrial body, estimates reserves of recoverable oil at about 350 billion barrels, a quantity sufficient to cover present levels of demand for sixty-five years. Potentially recoverable natural gas is estimated at 1.2 quadrillion cubic feet, sufficient to meet present levels of demand for more than fifty years. The Bureau of Mines estimates coal reserves at almost 400 billion short-tons, equal to a 700-year supply. The Atomic Energy

Commission "conservatively" estimates uranium reserves at 1 million tons of U235, sufficient to cover total electric-power needs at present levels for twenty-five years.

What exists, therefore, is an abundance of energy supplies but a shortage of energy actually available or acceptable in terms of price, environmental effect, geographic and political considerations, and technological capability. Overcoming these obstacles will take time, money, and technology. Possibly more important, it will take foresight, common sense, and goodwill.

Thus the short-term outlook is not nearly as sanguine as the earth's long-term capabilities would suggest. "Energy crisis" is in fact the most accurate term for the present situation. Those who criticize the term will find little support by referring to the dictionary. There *crisis* is defined as "the point of time where it is decided whether any affair or course of action must go on; or be modified or terminated; the decisive moment, the turning point." The energy situation in the United States in the early 1970s clearly fits this definition.

A troublesome problem now, energy policy could become a major social issue in the future, as people find their homes cold and their factories in short supply. Indeed, activists, including some consumer and environmental groups, are making energy a rallying point, a sort of peacetime Vietnam. Their reasoning is that it will be both a long-term and high-profile issue.

In the realm of establishment politics, congressmen have aligned themselves on the various sides of the energy issue. While ten years ago only legislators from the oil states took a deep interest in energy matters, now almost every congressman feels obliged to speak out on the subject. Local-government officials also feel duty bound to become energy experts. Governors, mayors, and even city-council members are getting involved.

The energy crisis has created a new form of schizophrenia in certain political circles. Politicians who have long been against support for domestic oil production are now strong advocates of United States self-sufficiency because of the effect that American dependence on Arab oil would have on the United States's relationship with Israel.

Part of the confusion results from the multiplicity of energy crises. The problems of the United States differ from those of the rest of the world; but in all cases solutions must be sought in short, intermediate, and long terms. The interlocking nature of the world energy system makes some ideal short-term solutions disastrous in the long run.

Similarly, actions that might be beneficial to the United States could well be unsettling for other nations.

For mankind as a whole the basic danger will not come until well into the next century when man has consumed most of the earth's storehouse of usable fossil fuels. Alternate sources, such as solar or nuclear energy, will have to be found and perfected or the human race could return to caves. For the United States the most pressing problems are short term. Even before the Arab embargo America had already begun to experience them in the form of blackouts, brownouts, and shortages of natural gas, fuel oil, and gasoline.

A major part of the short-term problem is logistical in that there appears to be little chance that domestic oil fields, refineries, and conservation programs can be made effective in time for the country to avoid difficulties of one sort or another. It takes three years to build a refinery, five years to create a port, five years to develop an oil field, and eight to ten years to build a nuclear plant. John Lichtblau, head of the Petroleum Industry Research Foundation, has observed: "If an oil field as large as Prudhoe Bay in Alaska, the largest ever discovered in the history of North America were found tomorrow in New Jersey, it would be a big help but the United States would still have energy problems for the rest of the decade."

The energy crisis covers several distinct energy-related problems. It is basically an immediate problem in that if the right decisions are made now, there should be no problems over the long term.

Progress is being made. Additions to the domestic refining capacity are being built; the Alaska-pipeline issue may soon be resolved; the federal budget for alternate energy sources has been greatly enlarged; and a new superagency to try to bring reason to energy matters has been created. Yet, as one Washington wag said, "President Nixon's April energy message was too late before it was delivered. To have been effective it should have been President Johnson's energy message and sent to Congress seven years ago."

How serious is the nation's energy situation? Senator Henry Jackson, considered by many to be the most informed elected official in Washington on the subject, describes it as "the most difficult problem facing the nation either internationally or domestically." This assessment is certainly open to debate, yet it can be argued that the nation's monetary, inflationary, environmental, and employment problems cannot be solved if the energy system that runs the economy is out of order. Urban and racial woes will only be exacerbated in a poorly functioning economy. The effect of a staggering, stumbling, energy-short America on the political stability of the world could be disastrous.

The United States has arrived at its present condition through a combination of growing demand for energy, declining fuel reserves, inadequate planning by government and industry, and waste of non-replaceable resources.

There would seem to be enough blame to divide responsibility for the problem among the oil industry, consumer advocates, environmentalists, and the public, which wastes more energy in a day than Japan consumes in the same time.

The one thing that everyone seems to agree on is that the chief culprit is the government, which, as a result of pressures from various groups as well as lack of foresight, has failed to create a stable, well-thought-out energy program for the nation.

Instead, what passed for a national energy policy was a jerry-built system that has crumbled under internal and external pressures. Each attempt to patch up the ungainly mechanism resulted in another segment becoming unglued unexpectedly. The question of how the United States reached the stage where even the slightest possibility of such a dire future could exist will most certainly be a topic of study by historians as it is already for politicians. S. David Freeman traces the cause to the days of the New Deal when the government instituted the policy of keeping energy costs as low as possible in order to stimulate ever-expanding use. He believes that although this policy was right for its time, it was maintained too long and has led to the present shortage.

Two basic schools of thought on the matter have developed, which, for the sake of convenience, can be identified as Malthusian and Machiavellian. The Malthusians are mostly independent economists, oil-industry executives, bankers, engineers, geologists, and government officials. They believe that the present situation is the result of natural consumption of a finite resource (fossil fuels) by growing demand that has been accelerated and distorted by political events. Peter Peterson, the former secretary of commerce, typified the Malthusian outlook when he described the nation's energy situation as a case of "Popeye running out of cheap spinach."

The Machiavellians tend to be academic economists, liberal congressmen, members of consumer-interest groups, and activist lawyers. They view the present energy situation as a contrived affair orchestrated by the oil interests to squeeze more money out of the public. A comment by Thomas J. McIntyre of New Hampshire carries the general message of the Machiavellians. He said, "Either the Federal officials responsible for oil policy in this country displayed an unbelievable level of incompetency or the petroleum industry itself misrepresented the facts. I personally believe that a combination of both factors were at work."

The Machiavellians contend that for years there has been too much collusion between the government and the industry. They point out that the bulk of statistics and data used by the government is supplied by the industry and that the National Petroleum Council, which advises the government on energy matters, is composed of oil- and gas-company executives.

Many Machiavellians date the beginning of America's energy problem from the imposition of mandatory oil-import quotas by President Eisenhower in 1959. The rationale of the program was to create a balance between foreign and domestic oil that would allow the United States to defend itself from military and economic threats. A strong undercurrent, however, was the demands of domestic oil producers for protection from cheap foreign oil.

Economists of a Machiavellian bent contend that quotas distorted the natural market mechanism, creating an inefficient national energy system. Consumer advocates have called the import system "a rape of the consumer." Helen Nelson, president of the Consumer Federation of America, has commented: "Since the inception of the oil import program, American consumers have paid some $50 billion in higher oil and gasoline costs. And what have they gotten in return for $50 billion? Shortages. It is no exaggeration to say that rarely in history have so many paid so much for so little."

Opposition to the import program was muted until 1967, when an attempt to create a foreign-trade zone and refinery in Machiasport, Maine, pitted the industry against the weight of the entire New England Congressional Delegation. The New England congressmen to this day have neither forgiven nor forgotten the opposition. Many oil men privately admit that they overreacted on Machiasport and would now be glad to get permission for a refinery anywhere on the East Coast.

The Malthusians date the beginning of the energy crisis from 1954 when the Supreme Court, in clarifying the Natural Gas Acts of 1938, ruled that the Federal Power Commission had the right to control producer wellhead prices. Thornton Bradshaw, president of the Atlantic Richfield Corporation, typifies the Malthusian view on the issue:

Since that time, the F.P.C. has struggled with the impossible task of regulating the price of a commodity produced by thousands of competing units in a high risk business. Because no economic standards were available the Commission has sought refuge in a politically expedient standard—the lowest possible price to the consumer. This had the result foreseen by most economists—demand soared, because gas was cheap, supply declined because incentive no longer existed to search for gas.

The Malthusians note that the equivalent amount of energy from oil costs 60 percent more than from gas and point out that since the decision, the number of gas wells drilled has fallen sharply, as have discoveries.

It would seem that both oil-import controls and the decision to regulate gas prices were well intentioned but poorly thought out and even more poorly enforced. The quota system for oil imports has been eliminated, and it is likely that the regulation of natural gas is being changed. But canceling out the actual issues will probably do little to close the philosophical chasm between the two groups, which could actually widen if and when the energy crisis becomes critical.

The most critical immediate problem facing the country concerns natural gas. Over the last three years, schools and factories in various sections of the nation have had to close for short periods for lack of gas. The Federal Power Commission reports that curtailments of natural gas by suppliers to customers with firm contracts are increasing from 350 billion cubic feet in 1971 to 910 billion cubic feet in 1972. Curtailments from November 1972 to March 1973 are estimated at about 480 billion cubic feet. Natural gas is the cleanest and therefore most environmentally desirable of the world's traditional fuels. But production in the United States is expected to peak in 1975 at 24.7 trillion cubic feet. Production has exceeded additions to reserves each year since 1968.

The industry contends that higher prices will indirectly increase supplies by increasing incentives for exploration, but how much of an increase would be involved is a question. Moreover, a price rise would have little immediate effect. In the next year or two, more plants and schools will probably have supplies cut off.

Coal appears to be the obvious answer to the nation's energy problems. It is by far the most abundant source of fossil fuel, with several hundred years' potential supply. But, as S. David Freeman, head of the

Ford Foundation's energy study, put it, "There are two things wrong with coal today, we can't mine it and we can't burn it."

Coal for all its potential is a difficult fuel. Obtaining it means deep mining or strip mining, which are objectionable because of human hazards or environmental depredation. Moreover, consuming coal carries its own environmental threat since about 80 percent of supplies are high in sulfur content and emit harmful pollutants into the air as they are burned. Coal is also expensive and unwieldy to transport.

Efforts are being made to clean the pollutants out of the smoke pouring from the stacks at large, coal-burning utilities and factories. The technology, however, has not yet been perfected. High-sulfur coal can also be processed into clean natural gas or oil, but the technology for this transformation, though available, is not economically feasible.

Nuclear energy is now showing signs of living up to its promise. The Atomic Energy Commission predicts that nuclear-power capacity will represent about 23 percent of the nation's total electric capacity by 1980, 32 percent by 1985, and about 50 percent by the year 2000. The National Petroleum Council expects nuclear power to account for about 17 percent of the source by 1985.

America's nuclear trump is the fastbreeder reactor, which theoretically can generate more nuclear fuel than it consumes over a thirty-year period, creating a sort of perpetual-motion energy machine. But it has drawbacks too. The first "commercial demonstration" unit in the United States is not scheduled for operation until 1980, and the breeder creates massive amounts of radioactive wastes, whose danger to man lingers for thousands of years.

An even more exotic nuclear solution is fusion, the process that powers the hydrogen bomb and is the energy source for the sun and the stars. Its basic fuel would be hydrogen, which is so abundant in sea water that it would provide enough energy to meet the world's electricity needs for as long as anyone can project. But it will not be a factor, if at all, until the twenty-first century.

It is likely that oil, the present mainstay of American energy, will have to do even more than in the past. Oil is the most versatile energy source. It is a swing fuel capable of running huge electrical generators or the family car. Just as important, it is easily transported and can be used in hundreds of products from gasoline to plastics. As a result, oil is expected to be carrying more than 50 percent of America's energy burden by 1980.

The trouble is that the United States exhausted its reserve-production capacity in 1970 and will probably never again be able to meet its oil needs from domestic sources. In 1973, demand will exceed 17 million barrels a day, with domestic production at little more than 11 million barrels a day. By 1980, demand will be around 24 million barrels a day with production at about 12 million, according to the State Department.

As for solutions to the problems, most experts say there are four basic approaches: sharply increased oil and gas imports; greater exploration and production from domestic areas; development of alternate energy sources; and energy conservation programs. Each of these approaches, however, has its problems. Increased imports bring with them dependence upon potentially unreliable foreign sources as well as an increase in the United States's balance-of-payments deficit that is likely to be massive and unacceptable. Greater domestic exploration makes higher prices and environmental hazards almost a certainty. Alternate sources, such as nuclear fusion or solar or geothermal energy, are unproven, very costly, or far in the future. Conservation programs, if enforced strictly, could raise a political furor, and if applied loosely, could make only a minor contribution. The administration's policy before the Arab embargo was clearly one of buying time through imports while it attempted to stabilize the situation through increased domestic production, new energy sources, and conservation.

In 1965, the United States imported 2.5 million barrels a day, or about 20 percent of its demand, with all but 2 percent coming from Venezuela or Canada. Imports inched up to 23 percent by 1970, but jumped to 27 percent in 1971, and in 1973 reached about 37 percent of total demand. By 1980, the United States is expected to depend on foreign sources for between 40 and 60 percent of its oil. This country's traditional suppliers, Canada and Venezuela, appear unable to increase their output significantly, so that much of the new imports must certainly come from the Middle East, where 75 percent of the world's proven oil reserves lie buried beneath the sands.

The failure to discover a substitute for Middle East oil is not from want of trying. In recent years 95 percent of new exploration activities by the oil companies has been outside the Persian Gulf. America's dependence on foreign energy sources has come at a time when the balance of power in the international oil trade has shifted decidedly to the producing nations after residing with the oil companies and thus with the consuming nations since the inception of the oil trade in the early part of this century.

Now the producers are calling the shots. Arab political pleas, so long ignored, have turned to demands—demands that Europe and Japan are obeying and the United States would probably agree to, were it not for domestic political considerations. Producing-government revenues from a barrel of oil have doubled in just one year and most analysts anticipate continuing price advances. The developed, consuming countries are in disarray, not the undeveloped oil-producing countries.

An unfortunate near certainty in the world picture is the future of the developing nations of Asia, Africa, and South America. These countries will have to pay sharply higher prices for both energy and manufactured goods. Development will thus be more difficult.

Some analysts contend that energy difficulties are just the first signs of coming worldwide shortages in other basic commodities. Indeed, the International Institute for Strategic Studies recently commented: "It was clear from Soviet-American exchanges or promises of exchanges of food and fuels on a massive scale that raw material resources were becoming once more a major potential factor in the world political balance." The United States is self-sufficient in only ten of the thirty-six basic industrial raw materials and the Soviet Union in only nine.

Some sobering thoughts on the future come from M. King Hubbert, research geophysicist for the United States Geological Survey, who notes, "What stands out most clearly is that our present phase of exponential growth, based upon man's ability to control ever larger quantities of energy, can only be a temporary period of about three centuries duration in the totality of human history. It represents but a brief transitional epoch between two very much longer periods, each characterized by rates of change so slow as to be regarded essentially as a period of non-growth."

Lessening the Impact

Energy Conservation

ALLEN L. HAMMOND

Everyone is for energy conservation so long as they do not have to change their life-style, give up their oversized car, or pay more for a well-insulated house. A prime cause of present energy problems, in fact, is a spendthrift attitude that has led to inefficiency and just plain carelessness. Waste has become institutionalized in building codes, in the design of many commercial products, and in industrial manufacturing processes. The magnitude of this waste is appalling. Five-sixths of the energy used in transportation, two-thirds of the fuel consumed to generate electricty, and nearly one-third of the remaining energy—totaling more than 50 percent of the energy consumed in the United States—is discarded as waste heat. With potential shortages of energy threatening economic stability and the general welfare of the country, energy conservation is likely to become an essential national goal.

Energy conservation is far from a new ethic. Earlier generations in this country did not waste firewood when they had to cut it themselves. Nor were habits such as turning out the lights when leaving a room just a manifestation of puritan values. Until the last decades electricity was very expensive. But the American industrial miracle, especially in the period following World War II, made available greater supplies of oil, natural gas, and electricity at lower prices. The price of electricity, for example, when measured in constant dollars (that is, adjusted for inflation), decreased every year for twenty-five years after 1946. Energy became a better and better buy, everybody used more, and no one looked ahead. In both public and private decision-making, it was taken for granted that adequate supplies of cheap energy would always be available.

Today it is clear that this assumption was in error and that many public and private policies were shortsighted. It is perhaps not too strong to say that the growing shortages of traditional fuels and the rising prices for all forms of energy mark a new era of energy use. Energy, the basic natural resource and the basic industrial good, will cost more, perhaps much more, than in the past. Some economists have speculated that by the end of the century the cost of electricity might double, the cost of natural gas might triple, and the costs of uranium, coal, and petroleum might increase substantially.

The energy crisis has been described as a fuel crisis, an environmental crisis, and a management crisis. Although new sources of energy will help, their development will take time and money.[1] Making existing supplies of fuel go farther will buy time. Slowing the rate of growth of energy use will also alleviate the environmental problems associated with energy production and use. Even without wholesale changes in consumer habits, there are possible conservation measures, many of which involve little or no change in life-styles. Technological changes and more rational policies could markedly ease the energy shortage in the coming decades.

In 1971 Americans consumed energy at the rate of about 2 billion kilowatts, the equivalent of more than 2,000 large power plants running at full capacity—more than twice as much as twenty years ago. About 96 percent of the energy used in the United States comes from fossil fuels: petroleum, 43 percent, mostly for transportation; natural gas, 33 percent; and coal, 20 percent. Hydroelectric energy accounts for about 3 percent of present production, and nuclear energy for about 1 percent. Twenty-five percent of the raw-energy resources consumed were used to generate electricity, and this percentage is increasing rapidly. Transportation accounted for about 25 percent; industry for 30 percent; and residential and commercial consumption for the remaining 20 percent. Slightly more than half of the electricity generated was used by residences and commercial buildings; the remainder was used by industry.

[1] For a more detailed discussion of current and future energy sources, of environmental problems, and of economic prospects for new energy technologies, see Allen L. Hammond, William D. Metz, and Thomas H. Maugh II, *Energy and the Future* (Washington, D.C.: American Association for the Advancement of Science, 1973).

According to a study conducted by the Stanford Research Institute of Menlo Park, California, for the now-defunct White House Office of Science and Technology, the largest single application of energy is transportation fuels, primarily gasoline, which accounts for 25 percent of total energy consumption. Space heating in homes and commercial buildings consumes 19 percent, industrial applications of process steam 17 percent, industrial direct heat in furnaces, kilns, and the like 11 percent, and electric machinery 8 percent. Other end-uses are smaller but still significant: water heating 4 percent, air conditioning 2.5 percent, refrigeration 2 percent, and lighting 1.5 percent. The use of air conditioning is growing most rapidly, increasing more than twice as fast as total energy consumption. Even 1 percent of total consumption is equivalent to 100 million barrels of petroleum, and well worth some effort to save.

One prime category for conservation efforts, and in the long run perhaps the most significant, is homes and commercial buildings. Most structures have not been designed to conserve energy, and their thermal performance is often poor. The list of energy wastes includes inadequate insulation, leakage of outside air, excess ventilation, large window areas (especially in modern office buildings), inefficient heating and cooling equipment, and excessive lighting.

The Federal Housing Administration's guidelines for insulation in residences permitted heat losses as high as 50 BTUs per square foot of floor space per hour in 1965. Revised guidelines issued in 1971 reduced this figure but still do not require the economically optimum amount of insulation. According to calculations by John Moyers, a scientist at Oak Ridge National Laboratory in Tennessee, additional insulation in walls and ceilings, foil insulation in floors, weather stripping, and storm windows would save money as well as energy for homeowners in some parts of the country. In some cases Moyers called for nearly twice as much insulation as the federal guidelines. Because the guidelines are only voluntary, many private builders ignore them. Another problem is the fact that the guidelines apply only to new construction. Most houses fail to meet even the revised standards, and many older buildings have little or no insultaion. Although backfitting existing houses with more insulation is difficult, this measure alone would prevent shortages of heating oil this winter and would guarantee adequate supplies for some years to come. Upgraded and rigorously enforced standards in all new construction would eventually accomplish the same thing, but more slowly.

Commercial buildings are underinsulated too. Charles Berg of the National Bureau of Standards in Gaithersburg estimates that 40 percent of the energy used in heating these buildings could be saved. Modern high-rise office buildings in particular are among the worst offenders. Perhaps the classic example is New York's World Trade Center, which uses energy at the rate of 80,000 kilowatts, more than many medium-sized cities. Electric heating is especially inefficient in such buildings, since they could readily burn fossil fuels and even wastes in large, well-controlled furnaces. Other energy savings are possible with reflective window glass (or just less glass), reduced air ventilation, and air-conditioning plants that operate on heat, not electricity.

Architectural practices often promote excess energy use. Richard G. Stein, an architect in New York City, points out that poor design often results in the overuse of steel, concrete, and other energy-intensive materials by as much as 50 percent. Nearly a quarter of all electricity is used for lighting. While the illumination levels recommended in commercial buildings have more than tripled in the past fifteen years, there is now considerable disagreement about whether such high illumination levels do more harm than good. Particularly questionable practices are uniform-intensity light, leaving lights on all the time, and providing only a single switch per floor. Stein believes that as much as a 4-percent saving in total electricity use could be achieved immediately by reducing excess lighting in existing buildings and by more carefully designing new buildings. Overall, Stein estimates that as much as half of the energy used by modern office buildings could be saved. Reductions in peak power demand, which strain electrical systems and cause brownouts, would be even greater.

In addition to structural improvements in thermal design and the wiser use of light, more efficient heating and air-conditioning equipment could conserve energy. The efficiency of window-unit air conditioners sold today ranges from 1.37 to 3.57 (BTUs of cooling per BTU of energy consumed). If the average efficency of these units in the United States were upgraded to about 3, the energy saved would amount to some 16 billion kilowatt-hours per year. Again, peak power demands during summer months would be reduced significantly. At present, however, efficiencies are not usually stated explicitly on commercial units, nor do many consider that a slightly higher initial cost will be rapidly repaid by lower operating costs.

Home furnaces can be as efficient as 75 percent, with the rest of the heat lost in the exhaust. Mismatched capacities and infrequent maintenance often lead to efficiencies as low as 35 to 50 percent. Electric-resistance heating, now being installed in about a third of all new construction, is essentially 100-percent efficient in place, but every kilowatt-hour of electricity costs 3 kilowatt-hour equivalents of heat to produce, and about 10 percent of the electricity is lost in transmission and distribution. Consequently, electric heating is inherently less than 30-percent efficient.

Electrically driven heat pumps, however, could improve the efficiency of electric heating. Heat pumps are basically air conditioners run in reverse. Although not now widely used, they deliver an average of two units of thermal energy for each unit of electrical power consumed. In the past, frequent failures and high maintenance costs have been common, but improved models are now available. Heat pumps may become an increasingly attractive option for the homeowner as fossil-fuel costs rise and nuclear power plants become the prevailing source of electricity.

The ultimate space-conditioning methods would utilize solar energy, which, as a renewable resource, is also piped conveniently to the door. Commercially available solar-heating and -cooling equipment do not yet exist, although the National Science Foundation predicts that they may be available within a few years. Several dozen pioneers, however, have installed homemade systems. Solar hot-water heaters have been in limited commercial service for some time.

Still other approaches to minimizing energy consumption in the home have emerged from an ongoing study directed by D. G. Harvey of Hittman Associates in Columbia, Maryland. Harvey notes that outdoor gas lights and pilot lights are surprisingly large fuel-users and points out that electronic igniters are commercially available. Other major causes of energy loss, he found, were leaky window frames, open chimney flues in unused fireplaces, and frost-free refrigerators, which use almost twice as much energy. Suggestions for energy conservation include heat-recovery systems in furnace flues (a possible 12-percent improvement), fluorescent lights (a 75-percent improvement over incandescent lights), well-insulated ovens, including most self-cleaning ovens (not all labor-saving devices waste energy), and deciduous trees that shield the roof in summer but not in winter.

A little-noticed trend that may have an important effect on energy consumption is the increasing number of mobile homes, recreational vehicles, campers, and the like. Mobile homes alone account for one out of every four new dwellings built in the United States. Because of their thin walls, limited insulation, and box-like construction, these dwellings use large quantities of energy. Inefficient space heaters in winter and several window air-conditioning units in summer are common additions.

Industry is the largest energy consumer, but, with some exceptions, the efficient use of energy has not been a prominent concern of American industry until lately. The production of primary metals, basic chemicals, petroleum products, food, paper, glass, and concrete consumes and wastes a great deal of energy. This situation may change rapidly as fuel prices rise. In fact, industry is the best hope for significant short-term energy savings. New vacuum furnaces have been developed that require only one-fourth the energy of earlier designs. Heat-recovery devices and better thermal management of many processes may also save considerable amounts of energy. Some investigators estimate that as much as 30 percent of the energy now used by industry might be saved—a goal that will test corporate management's vaunted flexibility in the face of changing conditions.

A case of particular interest is the utility industry, which has improved the efficiency with which electricity is generated from 5 percent in 1900 to nearly 40 percent in the newest coal-fired plants. The average for all existing power plants is about 32 percent. Coincidentally, nuclear power plants with light-water reactors (the type now common) also have efficiencies of about 32 percent. Proliferation of nuclear power plants is therefore not likely to improve overall efficiencies. However, power plants with high-temperature gas reactors are now becoming available, and these have efficiencies near 40 percent. The development of combined-cycle power plants, with gas turbines or magnetohydrodynamic generators in addition to steam turbines, could increase generating efficiencies to 50 or 60 percent. Widespread application of these improvements will take some time. Meanwhile the use of electrical power where fossil fuels could be used is clearly wasteful.

Although transportation constitutes the largest single use of energy, it offers few opportunities for significant savings. Because changes to more efficient modes of travel involve greater changes of life-style than do the other conservation measures discussed, these changes are harder to bring about. Nonetheless, the opportunities are striking. A study by E. Hirst of Oak Ridge National Laboratory reveals that during the 1960s passenger traffic on United States railroads decreased by one-half, automobile mileage increased by one-half, and airline mileage increased nearly threefold. Yet airplanes are only 40 percent as efficient per passenger mile for intercity transport as automobiles and 34 percent as efficient as trains. Within the city, automobiles use twice as much energy per passenger mile as buses, 27 times as much as walking, and 40 times as much as bicycling, the ultimate mode of transport as far as energy conservation is concerned. The disparities for freight transport are just as great. In energy per ton-mile, pipelines are the most efficient, railroads 1.5 times as costly, barges and other water transport about the same, trucks 8.4 times as inefficient, and air freight 93 times as inefficient.

Despite the comparative efficiency of foot travel and bicycle, most people, of course, prefer to travel by car. Indeed, Hirst estimates that when both direct and indirect costs are included, the automobile accounts for 21 percent of total energy consumed in the United States. The standard American car is much heavier than its European counterpart and gets only twelve miles per gallon on the average, compared with more than twenty miles for the foreign compacts. The poor performance of the American car is not due in any great part to pollution-control devices but rather to its size and weight.

In part, the decline of rail service and urban mass transit reflects the greater convenience, flexibility, and speed of automobiles and airplanes. But governmental promotion of automobile, truck, and airplane traffic through subsidies of roads and airports has also been influential. Reversing these shifts even in part could save significant amounts of energy, predominantly in the form of petroleum, and reduce the need to import this commodity.

More efficient use could easily save a quarter of the energy now consumed. Further savings could be made by reducing the demand. In an ideal capitalist economy, rising demand would cause higher prices, which would in turn reduce the demand. But for a variety of reasons, including government regulations and industrial monopolies, energy supply and demand do not constitute an efficient market in the economist's sense of that term. Because too many users get too much energy at too low a price, there is a shortage and a condition that economists describe as market failure. It is not clear, however, that higher prices (and higher profits for the energy industry) are in the consumer's best interest. Much of the growth in demand for energy is stimulated by advertising, by promotional rates for energy, and by a host of other policies and practices that might better be ended. Slowing the rate of growth in the demand for energy would not only provide time for the development of new energy sources; it would also mitigate the environmental problems caused by energy production and use, relieve the supply problems caused by shrinking domestic reserves of gas and oil, and reduce the international economic penalties of importing ever-larger amounts of these fuels. Thus, damping the demand for energy is beginning to receive serious consideration in federal and state governments.

Of course, the simplest way of limiting demand is to allow prices to rise, which they will, even in an inefficient and partially controlled market. These increases may have a dramatic effect on demand. For example, the Federal Power Commission (FPC) estimates that the demand for electricity will double by 1980 and almost double again by 1990. However, the FPC estimate and others like it are based on extrapolations of previous trends in overall economic and population growth and do not take into account rising prices. More sophisticated econometric studies of the demand for energy, such as that conducted by Duane Chapman and his colleagues at Cornell University, indicate that variations in the price of electricity and, to a lesser extent, in population growth can vary the projected demand for electricity in the year 2000 by fivefold. Noting that electricity prices have already started to rise and that the birthrate has fallen to less than the "zero population growth" replacement rate, Chapman finds that by the year 2000 demand will increase only 33 percent over the 1970 figure, assuming electricity prices double. If electricity prices increase only slightly, Chapman projects demand will be about 2.5 times the 1970 figure at the end of the century.

Other investigators of the influence of prices on the demand for power have found qualitatively similar results. Analysts at the Rand Corporation in Santa Monica, California, studied electricity demand for that state. They found that even with a booming economy, demand is not likely to grow nearly as fast as utility projections indicate and that increasing energy prices might well reduce electricity needs by the end of the century to half of the industry estimate.

Broad price increases will help to reduce demand, but larger and more rapid reductions are possible. Many observers advocate more specific strategies to eliminate particularly wasteful or socially undesirable uses of energy. One method of reducing the demand is to eliminate rate structures that provide incentives for high-volume consumers. In most states, big users receive such large discounts on the price of electricity that small users effectively subsidize the larger consumers. Because the industries and commercial interests that benefit from this discount would be the first to reduce their use of power when faced with higher prices, the method appears to be a particularly suitable one. The Environmental Defense Fund is intervening in a hearing before the Wisconsin state utility commission concerning proposed increases in electricity rates. Based on economic studies, the environmental group believes that by pricing power in proportion to the true cost of supplying each customer, Wisconsin could reduce the demand for electricity enough to eliminate the need for several projected power plants. The group plans to challenge the rate structures in other states.

Others have proposed more drastic changes in the cost of energy and in the tax system to retard the growth in the demand for energy and the exploitation of energy resources. Walter Heller, a former presidential adviser, now with the University of Minnesota, believes that large depletion allowances, capital-gains shelters, and special tax deductions should no longer be allowed for energy-producing industries. According to Heller, "the public is subsidizing these industries at least twice —once by rich tax bounties and once by cost-free or below-cost discharge of waste and heat."[2] The principle put forward by Heller and other critics of the existing financial incentives is that the cost of energy should reflect not only its scarcity but also the environmental and social costs of producing it.

Other strategies for energy conservation involve tax incentives to encourage more efficient building design, more efficient appliances, and the use of renewable resources such as solar energy. Regulatory measures could accomplish these same ends. Federal guidelines and state building codes could be changed to require more insulation and heat-reflecting glass in office buildings. Direct public investment, such as funding mass-transit systems or retrofitting whole subdivisions with additional insulation, is another option. Public education could provide the consumer with information about energy conservation, or promotional advertising for energy (by electric utilities, for example) could be deducted from profits rather than included in business expenses. In practice, both regulatory and market strategies will probably be of use. Of particular importance are changes in government policies that promote energy consumption or that, as in the case of natural-gas regulation, result in unrealistically low prices.

Just how feasible are energy-conservation policies, and how much energy might they save? Although little research has been directed

---

[2] Walter Heller, "Coming to Terms with Growth and the Environment," in *Energy, Economic Growth, and the Environment*, ed. Samuel H. Schurr (Baltimore: Johns Hopkins Press, 1972).

toward these questions, the evidence is that significant amounts of energy could be saved and the rest used more wisely. A study by the Rand Corporation, for example, found that by the year 2000, the electricity needs of California might be reduced (compared to what would otherwise be needed) by as much as 430 billion kilowatt-hours annually through conservation policies. These savings would reduce the projected need for new power plants from an estimated 127 to 45 or less and thus reduce environmental damage due to power generation. The conservation policies would induce only minor economic dislocations, the study concluded, and would not affect the growth of the state's economy. A separate study, by the Environmental Protection Agency, was national in scope and focused on energy saving before 1990. It found that potential savings included 22 percent of residential and commercial energy use, 34 percent of industrial consumption, and 27 percent of transportation-energy use, in all a reduction of nearly 30 percent.

It appears that there are ample means of increasing the efficiency of energy use and a variety of possible policies to accomplish this goal. How much energy is needed, doomsday headlines to the contrary, is very much a figure that can be large or not so large depending on what actions are taken now and in ensuing years. Higher prices will be one incentive to use less energy and to use it more efficiently. More direct measures to eliminate wasteful practices and to slow the growth in the demand for energy can be even more effective and not inflict as heavy a burden on the poor, who would bear the brunt of higher energy prices. Energy shortages have appeared in the United States because its technological skills and economic policies have not been adequately applied to the energy system. There are plentiful sources of energy and environmentally acceptable ways to produce it, but conservation must be an integral part of efforts to resolve present energy difficulties and more serious problems and shortages in the future. As indicated at the beginning of this article, conservation is not just a matter for the federal and state governments and for industry and commercial enterprises. For the most selfish of reasons, it is also a matter for every individual faced with gasoline shortages, higher electric bills, and pollution from inefficient use of energy in the transportation system and in generating plants. Energy conservation begins at home.

Prospects for Nuclear Power

JAMES E. CONNOR

For a generation, scientists and publicists have had to resort to the title "the prospects for nuclear power" because the nuclear present was hardly large enough to merit a short paragraph, let alone a lengthy article. Given this propensity to live in the future, anyone wishing to talk about the prospects for nuclear power ought to be required first to explain either why his own insight into the future is clearer than that of his predecessors or at least why looking into the future now might be expected to yield a clearer picture than before. Since the writer is not quite immodest enough to claim greater insight, his justification must rest on the appropriateness of the moment.

From several different perspectives the present can be seen as a watershed in the development of the nuclear industry. For example, the amount of energy produced in the United States from commercial nuclear power plants does not yet equal that produced by burning wood. From this perspective, the nuclear business obviously must be classed as an infant industry. The infant, however, is a healthy one. Commercial nuclear power plants in being, under construction, or on order today are equivalent to the entire electrical-generating capacity of the United States in 1960. A decade from now, Americans in all sections of the country will be drawing a significant share of their electrical energy from nuclear power plants. The past, then, is small enough to be comprehensible and the future large enough to be intriguing.

An End and a Beginning

Today's nuclear industry is the product of three and a half decades of unusually well-documented history. Impelled by a letter from Einstein to Roosevelt, nuclear science grew from a laboratory curiosity to a multibillion-dollar industry in three years. It did so under the strictest secrecy and with no financial constraints. That it succeeded in accomplishing its military objective without scandal or delay is testimony to the extraordinary collection of scientific, technical, and managerial talents that were combined in the Manhattan Project.

After the war two unique governmental institutions, the Atomic Energy Commission and the Congressional Joint Committee on Atomic Energy, were established to oversee the nation's nuclear establishment and to chart its future course. In their respective governmental spheres each institution was granted a monopoly. When they worked together they blended legislative and executive functions to an unprecedented degree. The fruits of the collaboration are noteworthy by any standard: creation of the nation's nuclear deterrent, its nuclear navy, and its emerging civilian nuclear power industry. It should not be thought, however, that these accomplishments depended solely on legal monopolies or on political accommodation. Rather, they derived from the extremely high degree of professionalism, continuity, and commitment that characterized the men in both institutions.

The story of these men has been told and retold in biography and reminiscence. They were, to a remarkable extent, self-conscious individuals who knew full well that they were engaged in an enterprise which had awesome implications for both good and evil and which for better or worse would irrevocably alter the course of the future.

In trying to assess the ways in which the nuclear industry might develop over the next decade, it is important to recognize that these men will not be around to shape the future as they did the past. In both the public and the private sectors, men who have known and worked with one another since the days of the Manhattan Project are retiring. Those taking their places have different backgrounds, interests, and motivations. The homogeneity of experience and perspective that marked the nuclear pioneers does not distinguish their successors. An industry built on a foundation of shared experience and expectation will have to rely on a more commonplace and more diverse set of human motivations for its future growth.

As the men change, so will the nature of the challenge. The exciting tasks of uncovering the implications of a new science and of building a new industry from scratch have already been superseded by the mundane but essential concerns for reducing cost per kilowatt-hour, managing waste products, and ensuring component reliability. Moreover, the sheer magnitude of the commitment to nuclear power in the next decade ensures that the burden of performing these ordinary tasks will fall on ordinary men. Over the next ten years the utility industry will spend almost $90 billion for nuclear power plants and related systems. That figure translates into staggering demand for engineering time and construction labor. No small cadre of men, regardless of their brilliance or commitment, could be expected to supply that level of demand. During the next decade the nation's nuclear power system will be built and run in large measure by people trained as engineers, welders, managers, and pipefitters in the less demanding and less compelling context of nonnuclear activities. These men will constitute an elite only to the extent that contractors can bid the best men away from other large-scale construction and technology ventures. The nuclear industry's future performance will therefore resemble far more closely the nation's average industrial and technological performance than it has in the past.

The political system appears to have recognized this change in the nuclear industry's status. At this writing, the proposal to establish an Energy Research and Development Administration (ERDA), which would combine both nuclear and nonnuclear research and development responsibilities, seems quite likely to receive approval by the Congress. Along with ERDA, a separate Nuclear Energy Commission (NEC) will be established with the important but limited function of regulating the commercial nuclear power industry from the standpoint of public health and safety. In the executive branch, at least, the monopoly of nuclear responsibility by a single agency will come to an end. This change reflects not only on organizational point of view but also an important psychological shift in the way that decision-makers perceive nuclear power. They no longer see it as an arcane enterprise meriting special status in order to keep access to it tightly controlled. Rather, it appears more and more as simply one among many energy sources. It is still not clear whether organizational changes in the executive branch will be accompanied by congressional changes. Even if they are not, however, the uniquely intimate relationship between the Joint Committee and the AEC is not likely to be duplicated by ERDA and the numerous committees that will supervise it.

Problems of the Future

The personal and institutional framework on which the first three decades of the nuclear industry were built will not support the industry in its fourth decade. A new framework is of course emerging, but its outlines are not yet entirely clear. The main elements will depend on the way that the industry deals with three distinct kinds of problems— those that are common to any rapidly growing industry, peculiar to the utilities, or related to specific nuclear characteristics.

One common problem of any new and expanding industry is its inability to keep up with the demand. The symptoms, which are easy to identify, have already clearly manifested themselves in the nuclear industry during the last few years.

Construction difficulties have been badly underestimated, and initial time and cost schedules for plant projects no longer bear any relation to reality. While manufacturing capacity has been strained by the rush of demands, those skilled in nuclear-related work have been in short supply. The contribution that can realistically be expected from nuclear power during the next ten years will, to a great extent, depend on industry's ability to cope with rapid growth.

In retrospect it is clear that the initial cost and time schedules for nuclear power plants were wildly optismistic, given the industry's lack of experience with the technology. Recently, scheduling has become more realistic, and it appears that the worst delays may be a thing of the past. Moreover, the trend toward longer and longer construction periods seems to have leveled off. Utilities can now expect with a reasonable degree of confidence to have a nuclear plant on line about seven years after they place the order for the nuclear steam-supply system. But although the situation is no longer growing worse daily, neither has it shown signs of growing better. There should be some signs of a learning curve as the manufacturers of power systems become more familiar with the requirements placed on them. These effects, however, have not been visible for several reasons. First, the learning curve may have been obscured by increased project scopes and design changes. For example, since the first units were ordered, average plant size has increased by 90 percent. Utility schedules have had to be lengthened by over 40 percent to accommodate the delays attributable to increasing size and evolving technology.

A second reason has been the eagerness with which manufacturers have been willing to agree to (or even to suggest) design changes in order to sell their products. Each new power reactor has too often been designed from the ground up. In the future, both buyers and sellers will have to exercise more restraint if nuclear plants are to make their maximum contribution to meeting the nation's energy needs. Finally, the AEC must share responsibility for longer schedules. It, too, has suffered from growing pains. Until recently, its regulatory arm had not been sufficiently staffed and organized to deal with a rapidly expanding work load. In particular, the absence of a full set of detailed standards against which vendors could test their designs and regulators could check the vendors has caused serious problems. In lieu of standardization, plants have been custom designed and custom reviewed, a procedure that has obviously increased the time required for completing a plant and discouraged any learning-curve effects.

Plant size and type now seem to be stabilizing. Vendors and utilities have taken several significant steps toward agreeing upon standardized designs. This trend is being encouraged by the AEC, and the results of standardization, including pronounced learning-curve effects, should be visible within the next decade.

The move into the nuclear age has perhaps had its greatest institutional impact on the nation's utilities. For decades, the utilities have operated as rather simple,. straightforward business enterprises, ordering equipment and technical services from other firms and holding those firms wholly responsible for their products. Seldom was any technical capability retained by the utility itself. If a piece of equipment did not perform satisfactorily and corrective action was not taken expeditiously, the vendor could be taken to court. Utilities expected delivery of reliable, bug-free equipment and passed the safely predictable costs on to their customers.

The utilities could operate in this fashion because there were few uncertainties about the costs or reliability of coal- or oil-fired plants. In most cases it could be assumed that the increased costs of larger plants would be more than offset by economies of scale.

The introduction of nuclear power jolted utility managers out of this comfortable posture. Nuclear technology imposed intricate and time-consuming architectural and engineering requirements altogether new to utility executives. The inevitable problems of unforeseen delays and cost increases that accompany the introduction of a new technology have been particularly traumatic for the utilities, as much because of the unfamiliarity of the problems as because of their magnitude. Most unsettling of all, utility executives have discovered that they are responsible for these troublesome new plants. By law, the AEC licenses the utility itself—not the manufacturer—to build and assure safe operation of a nuclear plant.

Despite the problems of getting started, however, the utilities seem to have come a long way toward staffing themselves to meet the demands that the shift to nuclear power is placing on them. Men familiar with nuclear technology have been recruited from the federal government, the military, and the nuclear vendors. In addition, men who demonstrate an ability to deal with the complexities of a nuclear construction project cannot help but be identified as good prospects for promotion within the utility. This combination of external recruitment plus internal natural selection promises to ease some of the problems for utilities in the next decade.

Other problems, however, may be more intractable. The most noticeable problem is the rising capital cost of nuclear power. The causes for such increases are manifold, and they range from poor estimating and low labor productivity to an ever-increasing list of safety and environmental-protection devices required by the government in the interests of public health and safety. This escalation process lacks some of the constraints common to other industries, since utilities are able to pass such costs on to customers, who have little choice but to pay their bills or forgo power. Regulatory commissions can do little to lessen the impact since the cost increases are usually fully justifiable within their rate rules. Interfuel competition, which formerly served as a surrogate for actual competition, has not been able to perform that function recently because of the ballooning costs of fossil fuels and, perhaps more importantly, their uncertain supply. In sum, the factors that in the past served to dampen power-cost increases no longer seem to be working. If rising cost trends continue unabated, there must inevitably occur a sharp contraction in nuclear-power growth as a result of either lessened demand or the reemergence of attractive fossil-fuel alternatives.

Thus far the thrust of the argument has been that the nuclear industry is losing its singularity, is becoming more like other sectors of the economy and, therefore, more susceptible to the winds and tides of general economic activity than has been the case in the past. Notwithstanding this shift, however, it must be acknowledged that the industry also retains peculiarly nuclear aspects that also affect the course of its development. Surprisingly the two major nuclear considerations, public acceptance and effective regulation, are not technical questions at all.

One is psychological and the other institutional in nature.

There is no way to determine precisely the extent to which the public's attitudes toward nuclear power have been shaped by the fact of nuclear weapons. But one need not be precise in order to appreciate the impact that mental images of mushroom clouds have had on the development of nuclear power. The nation learned of nuclear energy under circumstances that made the deepest and most permanent impression. The technology that obliterated Hiroshima and Nagasaki was at once frightening because it destroyed so completely, miraculous because it seemed to bring peace so quickly, and mysterious because it was accomplished in complete secrecy through the use of scientific principles that were neither familiar nor easily understandable to the common man. Today, nearly thirty years later, elements of fear, wonder, and mystery are still present in public opinion.

Such persistence is especially remarkable in view of the way the public responded to space flight, the only scientific and technical achievement of comparable magnitude. In that case, ennui set in relatively quickly. Trips to the moon and long stays in orbit have become so commonplace that only the high drama of men in danger seems capable of refocusing the public's attention on this greatest of man's explorations.

In assessing the prospects for nuclear power, the reasons for such ambivalent public attitudes are less important than the fact of the attitudes themselves. The history of the nuclear era cannot be rewritten to eliminate fear or to dispel mystery. As a result, nuclear development over the next decade will take place in an environment of public opinion which, though not actively hostile, will not be supportive or even benevolently neutral. The public seems to be watching the emergence of commercial nuclear power cautiously yet with a degree of sophistication that is probably greater than either proponents or critics realize. Because of the residue of fear, it is doubtful that people will hail the arrival of a nuclear power plant in their vicinity during the next decade. On the other hand, assuming that nuclear plants maintain their high safety levels, the apocalyptic rhetoric of the more vehement nuclear critics is not likely to elicit much of a public response either. For the time being, opinion is in a delicate state of equilibrium. Inherent caution on the public's part is balanced by a recognition that power is necessary and a willingness to accept the nuclear method as preferable to other approaches. This equilibrium can of course change quickly in either direction. Drastic power shortages can elicit a "damn the torpedoes" attitude. Conversely, a serious nuclear-related accident could lend some factual support to the efforts of those calling for a moratorium on nuclear construction. In the absence of a distinct shift in either direction, however, public opinion does not appear to present directly either a major constraint on or a major impetus to nuclear power in the next decade.

Indirectly, however, public opinion may be considered the source of one of the knottiest problems facing the nuclear industry in the next decade, i.e., the development of an effective and efficient system of regulation that protects the public's health and safety without imposing exorbitant costs or delays on the public in the name of such protection.

As part of its regulatory function, the AEC oversees the design and other technical aspects of nuclear power plants in order to minimize the risk of a nuclear accident. Although the federal government does not exercise detailed control over conventional power plants, technological regulation as a method of protecting health and safety is an accepted federal activity. For example, the Federal Aviation Administration (FAA) is charged with ensuring that new private and commercial aircraft are designed and built to operate safely.

Notwithstanding analogous federal regulatory activities, the form of regulation practiced by the AEC has evolved in its own fashion. For example, the AEC's regulatory procedures resemble those of a court or an economic regulatory agency much more closely than those of the FAA. Extremely lengthy public hearings are held at two separate stages in the construction of a nuclear plant. An appellate hierarchy reaching to the commission itself has been established to review decisions in detail.

Public hearings have been part of the AEC's approach to regulation since the Atomic Energy Act of 1954 established a two-phase hearing process, one prior to initiation of plant construction and the second prior to plant operation. In light of the course taken by many of these hearings in recent years, it is hard to believe that they were once viewed primarily as informal occasions at which the citizens of a locale about to receive a nuclear plant could learn of its benefits and have dispelled any residual qualms. Instead the hearings have been marked by delay and divisiveness. Increasingly, they have taken on a judicial cast in which the utility presents its case before a stone-faced regulatory tribunal while the opposition attempts to raise doubts about the validity of the applicant's case. After a series of motions, cross examinations, and summations that look very much like what goes on in a courtroom, judgment is rendered. Regulatory proceedings are beginning to be resolved by "out-of-court settlements" of some or all of the issues between the contending parties.

Although the Atomic Energy Act required public hearings, it did not mandate that such hearings should follow an adjudicatory format. This course evolved as the AEC responded to increasingly sophisticated and effective attacks on nuclear power.

It is tempting to explain the appearance of quasi-judicial procedures by observing that lawyers, who seem to know no other way of behaving, represent all parties in the proceedings. Unfortunately, satisfying as this answer may be to nonlawyers, it mistakes effect for cause. In fact, the trend toward more formal adversary procedures owes more to the AEC's perception of public opinion than it does to the career choices of the participants. In order to deal with public fears of nuclear power, the process of licensing a plant not only had to be open and objective, it had to appear so as well. From the outset, the critics of nuclear power have appreciated and used this fact better than the AEC, the manufacturers of nuclear power plants, or the utilities. They have continually questioned the legitimacy of the regulatory system, alleging a conflict of interest arising from AEC's dual responsibilities as a developmental or "promotional" agency and as a regulator. Such a duality, the critics contended, was obviously not in the public interest since, consciously or unconsciously, concern for health and safety could be subordinated to baser promotional motives. The only acceptable form of nuclear regulation, they claimed, was one emanating from a wholly independent agency.

Curiously, the validity in principle of the charge seems never to have been questioned seriously by either the AEC or the Joint Committee. In public, at least, the nuclear community seems to have agreed that developmental and regulatory functions do not belong together and that the proper question to ask was when, not why, the split should occur.

This belief is especially remarkable in view of the fact that technological regulation, as in the example of the FAA, has traditionally been a function of operating departments charged with developmental or "promotional" responsibility. Independence, on the other hand, has been a charcateristic of economic regulatory agencies like the Civil Aeronautics Board and the Federal Power Commission. This obvious distinction, however, has never been pointed out in public debates on the proper organizational setting for nuclear regulation.

It is hard to say whether the AEC's reluctance to defend itself vigorously against conflict-ot-interest charges derived from conviction that the charges had merit or simply from a tactical assessment that, regardless of merit, the case would be a hard one to defend publicly. In any event, by appearing to concede the potential validity of the charge, the AEC foreordained the choice of adjudicatory hearings. For only by making its activities resemble the socially accepted standard of fairness, i.e., the judicial process, could an agency operating under the cloud of conflict of interest hope to present the appearance as well as the fact of objectivity.

Once started, of course, the judicial format is extremely difficult to alter. The participants develop a stake in the rules and, more importantly, any deviation seems to signal a loss of objectivity, thereby tarnishing the image of the process in the public mind. The regulatory system has thus become locked into cumbersome way of doing business that does not improve the quality of its decisions and is incapable of dealing effectively or efficiently with the demands that will be placed on it over the next decade.

The establishment of a separate Nuclear Energy Commission may provide the occasion for revamping the system of regulation. Since NEC will in fact be independent, it should not be as sensitive as the AEC to conflict-of-interest charges and thus may not feel compelled to follow established approaches to regulation. Moreover, because the new commission will not be distracted by programmatic demands, it may be able to wrestle with some basic policy issues that must be resolved before a coherent system of regulation can be established. These issues include questions as fundamental as how safe is safe enough and exactly what kind of warranty is implied when the commission issues a license to a nuclear power plant.

Although the creation of an independent commission shows promise of remedying the deficiencies in the present system, it must also be recognized that it carries with it some potentially serious problems. Briefly put, the major difficulty is how anything gets done in an agency where nobody has a stake in saying yes. From the standpoint of simple bureaucratic self-interest, the only way an NEC employee, or commissioner for that matter, can really get into trouble is to have been the one who approved something that later goes wrong. Under such circumstances the natural tendency is to delay approval as long as possible, to approve only what has been approved before, to impose extremely conservative operating procedures, or to require installation of layers of safety devices with questionable utility.

The potential for mischief in the system would be lessened if normal competitive factors acted to reduce costs or if those who were affected by regulation had a strong interest in resisting cost escalation. Unfortunately, as pointed out earlier, the utilities are accustomed to passing these costs on to their customers with, of course, a suitable add-on for return on investment.

The magnitude of the problem can be roughly calculated by assuming that the additional costs imposed by the system will range from 10 to 50 percent of the amount otherwise required for developing the nation's nuclear power system or, in other words, anywhere from $45 billion to $225 billion between now and the year 2000. These are monstrous sums, of course, and if imposed on the American public the nation would have to forgo many other investment opportunities.

Quantifying the costs of public-health and -safety measures has never been a particularly popular exercise. In an area dominated by a climate of moral fervor it has been considered bad form to ask whether a proposal to prevent harm will really work or, if it does, whether it is worth the cost. This attitude has been especially true of nuclear energy, which carries the added psychological burden of visions of Armageddon. But it is precisely in such an area that one must be sensitive to the trade-offs that always exist between levels of cost and degrees of safety. If these relationships are not recognized by the Nuclear Energy Commission, it is quite possible that the public health and safety will have been protected, but at a cost so high that the public interest will not have been well served.

Alternatives to Oil and Natural Gas

RICHARD L. GORDON

Now, as in the past, many are advocating the rapid development of alternatives to oil and gas. Their views are questionable for two reasons. First, they grossly underestimate the difficulties of developing new techniques. Only coal and nuclear power, for which utilization techniques are well established, can make a significant contribution in this decade. Second, they exaggerate the economic need for alternatives. Furthermore, nuclear power, coal, and newer approaches may not all be viable. Even if the price of gas and oil continues to rise, these sources may be considerably superior to coal, nuclear power, and the newer sources.

This article stresses the economic forces likely to influence the development of various energy sources. It begins with a review of the forces that limit rapid change in market patterns. Since the views presented here depend critically on assumptions about conventional oil and gas supplies, the relevant arguments are outlined. Then the special case of coal-based energy alternatives is examined as an illustration of the difficulties in securing competitive alternatives to oil and gas. The breeder reactor is discussed briefly prior to presenting overall conclusions.

Complex economies simply do not make rapid changes in fundamental processes such as their energy-consumption systems. Develop-

This article draws heavily on research supported by the National Science Foundation and Resources for the Future, Inc., but the views expressed here are solely the author's.

ing a major technology requires many years of testing and retesting. Then several more years are required to build a series of plants of increasingly greater size and sophistication. Thus it may take thirty years to design equipment suitable for commercial operation. Then constructing the commercial plants will take at least another five years. In any case, all consumers will not shift immediately to the new technology because they are restrained by their past investments. Although this restraint is sometimes interpreted as a sinister desire to protect their huge investments, it actually results from a lack of need. If one already has a perfectly good furnace, he is not likely to install a radically new system. The disincentive is often increased by the significant extra costs of installing new equipment in plants not designed for it.

Moreover, the various energy sources differ considerably in their ability to meet different energy uses. An obvious example is transportation in which engines using oil products are the most practical. However, most energy is consumed in stationary uses. In these markets, a variety of distinct applications ranges from small furnaces, heaters, and appliances in individual homes to giant plants that generate electricity. Many of the techniques for using energy are more efficiently employed in large-scale operations. Some technologies such as those that use nuclear energy are prohibitively expensive except in large electrical power plants. The problem for coal is less severe. While oil and gas flow easily and create no waste-disposal problems, coal is difficult to move and leaves an ash residue. Since it becomes increasingly more economical to deal with these problems as the size of the operation increases, the use of coal has tended to become concentrated among large-scale operations. Over half the coal produced in the United States is used to generate electricity; the next largest use is for coke employed for pig-iron manufacture—the only case in which the solid state of coal is an advantage.

In the next decade, options are limited to commercially available sources that involve the use of coal and nuclear power to generate electricity. In later years, more radically new technologies can be introduced. Those feasible for the next ten to twenty years include synthesis of oil and gas from coal, synthesis of crude oil from oil shales and tar sands, and the breeder reactors that produce more fuel than they use. All nuclear reactors use the reactive material in their fuel loads and convert some of the nonreactive material to plutonium. Breeders, however, produce more plutonium and convert more of the energy to electricity. By the next century, such vast energy sources as nuclear fusion, heat within the earth, and solar radiation may be employed economically on a large scale.

A critical problem with many of these new sources is that their employment would require a substantial restructuring of energy consumption because they are most suitable for large-scale use. Perhaps ultimately an economy could be shifted to one in which all final energy use would be in the form of electricity, or vast industrial complexes could be constructed around a central heat source. Such developments would take more time than a shift to synthetic oil or gas. The technological details are best left to such surveys as that of Hottel and Howard.[1] The present section concentrates on the problems of electrical power generation in the next decade.

A fundamental limitation is imposed by the lead time required for plant construction. Under favorable conditions a large electrical power plant can be built in less than five years. However, numerous problems both in the construction process and in regulatory procedures have produced concerns that completing a nuclear plant may take as long as a decade. Therefore, the next decade's nuclear capacity has already been determined. Much of the coal-use capability has also been set.

Just as it is impractical to change to nuclear power without building a new plant, it is also impractical to change fossil fuels. It is fairly simple and cheap to switch from coal to oil or gas, but shifting to coal is a difficult process. To move to coal, a plant needs large-size boilers and space for installing elaborate coal-storage and -handling facilities. Even if the plant has these facilities, the conversion will be substantially more expensive than a shift from coal to oil. In the past, electric utilities that expected to switch fuels initially built plants with a coal-use capability. As the probability of using coal diminished in certain regions, the electric companies ceased installing coal-use facilities and removed existing equipment.

Many environmental problems arise in electrical power generation. For coal, the critical problem is regulation of sulfur-oxide emissions. The use of coal in the electrical power industry has been concentrated in areas where the cheapest coal usually has a higher sulfur content than is acceptable under existing pollution standards.

While many ways to alleviate these problems were conceived, pri-

---

[1] H. C. Hottel and J. B. Howard, *New Energy Technology: Some Facts and Assessments* (Cambridge: The M.I.T. Press, 1971).

mary stress, at least for the 1970s, was placed on stack-gas scrubbing, which prevented the discharge of the sulfur oxides by cleaning the exhaust gases. This engineering task has taken much longer than early proponents of the approach had predicted. Despite these problems, some still insist that the systems are very close to commercial development and could be installed in a substantial portion of the coal-burning power plants in the United States by 1980. The electrical power industry, however, is much more skeptical about the prospects.

In any case, there is considerable interest in processes that remove sulfur before combustion. Because these processes require a chemical alteration of the coal, simpler and cheaper processes for manufacturing synthetic fuels are being developed. A good substitute for crude oil or natural gas is difficult to produce from coal because it is deficient in hydrogen. Therefore, it is necessary to upgrade coal gas by adding hydrogen. However, this upgrading is not essential for a power-plant fuel. The main options are to manufacture a low-BTU gas at the power plant or to use the solvent-refining process, which produces either a deashed solid or a liquid fuel. Sulfur could be removed during processing of both gas- and solvent-refined coal. Already various small-scale plants for both processes are under various phases of development. It is conceivable that a viable technology could be developed within a decade.

Similarly, electric ultilities have begun to use a combined cycle in which a gas turbine (a stationary adaptation of a jet engine) is connected to a conventional boiler that uses the waste heat from the gas turbine to power a steam turbine. While technical progress in conventional plants is difficult, the prospects for improvements in the combined cycle are quite good. The main drawback lies in present limitations on light fuel oils and gas, which are used in gas turbines. Attempts are being made to develop either an economical process for producing gas or a turbine that can use heavier fuel oils.

Nuclear power is presently under considerable attack for its environmental impacts. Some of the criticism seems ill-founded; the concerns about routine releases of radioactivity are based on dubious statistics. Besides, more effective controls can be developed. However, it has not been satisfactorily established that large accidental releases can be prevented. There are still issues of ensuring power-plant safety and of storing nuclear wastes for the millennia during which such material will be hazardous.

Conventional Oil and Gas Supplies

Advocates of energy alternatives validly point out that the growing world economy will some day exhaust the supplies of oil and gas unless other energy sources are substituted. The critical question is the urgency of developing such alternatives. The typical scenario postulates a need so pressing that massive research must be started yesterday. Another view is that options such as nuclear fusion can be developed more leisurely.

Amid the bitter debates presently raging, there is surprising agreement that world oil supplies can meet demands for at least a decade without significant rises in production costs. It is similarly recognized that the actual price of much of this oil will be affected by the efforts of a group of major producing countries to secure greater incomes by raising prices. Middle East oil was introduced into world markets at prices just low enough to displace Western Hemisphere oil from European markets. A long and fitful process lasting until 1971 brought these prices down substantially but left them well above costs. In 1971 the producing countries imposed high taxes that oil companies could pay only by raising prices. These states promise to continue increasing taxes.

The dispute about future oil supplies centers upon the efforts of the producing countries. However, this discord may be smaller than generally thought. Economic history, including that of the oil industry, suggests that concerted efforts to limit competition are difficult to maintain, particularly if consumers are clever about exploiting inherent conflicts among suppliers. Those optimistic about oil prices believe that these considerations will ultimately undermine the producing governments' cartel. Some may argue that this cartel is so unique that it cannot be overcome; however, others believe that consuming countries could formulate effective countermeasures but lack the unity to implement such policies.[2]

These optimists are right that consuming nations could greatly benefit from better policies. The only way to overcome the present barriers to consumer-country cooperation is to make the resulting benefits more apparent. Whatever the merits of the pessimistic view, it is too vague to deter the adoption of alternatives. Substantial oil and gas

---

[2] See the 1973 Senate Interior and Insular Affairs Committee hearings, *Oil and Gas Import Issues,* which includes a summary of the optimistic view by M. A. Adelman and an outline of the pessimistic view by James E. Atkins.

supplies are outside the cartel's control. At one time, it was believed that oil and gas discoveries in Alaska and the Canadian Arctic might eventually liberate North America from the need to import oil. The long delays in developing Alaskan oil have apparently not eliminated this long-range potential. The vision has simply been suppressed by the curious agreement of the otherwise bitterly opposed environmentalists and oil companies. Downgrading Alaskan sources makes opposition by environmentalists seem less damaging; the conservatism about Alaska is also useful to oil companies in building support for other projects.

From this discussion, the very least that should be concluded is that energy prices will probably not be high enough to make many alternatives viable within a generation. It can be contended that more likely oil will remain the world's basic fuel throughout this century. However, this view is not critical to the present argument.

Coal as an Energy Alternative

For generations, observers have misused data on coal reserves to argue that coal is the only fuel capable of meeting long-term world energy supplies. Actually, the only valid inference that can be drawn from coal-reserve data is that they are compiled quite differently from those for other fuels. Oil and gas reserves are computed only after the oil has been found, facilities for its production have been installed, and engineers have carefully measured the portion of its contents recoverable at prevailing prices. Coal reserves are based on geological surveys in which the indications provided by surface outcrops are extrapolated to provide estimates of total physical supplies. Thus the validity of these figures is doubtful even as measures of physical endowments. Moreover, the compilation pays little attention to the economic viability of these coal resources. It is not known how much cheaper coal really would be to mine and use than even the most costly convential sources of oil and gas. Therefore, it is uncertain that coal really can expand its markets in the foreseeable future. Indeed, there is a distinct danger that coal may also lose its present markets.

Traditionally, coal production has been centered east of the Mississippi. The main production area is Appalachia from Pennsylvania to Alabama. West Virginia, Pennsylvania, Ohio, and eastern Kentucky are the principal producers. A second major coal region encompasses Illinois, Indiana, and western Kentucky. While substantial reserves also exist in the Rocky Mountain states, a lack of markets has long limited coal production in this region.

A number of important differences prevail between eastern and western coal mining. Western coals are typically lower in sulfur content. Many western coals can be burned without treatment and still meet air-pollution regulations; most eastern coals are too high in sulfur content to meet such requirements. Generally, increased western coal production can more easily be secured through strip mining, although in some specific mining areas—especially Colorado and Utah—strippable reserves are limited. East of the Mississippi the potential for strip mining has never been great enough to permit complete elimination of underground mining. In addition, coal- and electrical-power industry sources indicate that most of the best strippable reserves in the East have been committed to mining. However, in many of the western states substantial uncommitted strippable reserves remain.

The main drawback of western coal is that it is expensive to transport to market. It is located a thousand miles or more from major centers such as Chicago. To make matters worse, the coal is usually a lower grade than eastern coal so that a larger tonnage volume is required to replace eastern coal. For example, to ship a ton of western coal to Chicago costs at least $6, which is about thirty-five cents a million BTUs. Coal could be shipped from downstate Illinois for about ten cents a million BTUs.

The difference in mining conditions has a profound impact on the comparative economics of the two sections. It is unlikely that eastern strip-mining costs for new mines will be less than those for underground mining. If strip mining could meet demands at lower costs, it would have done so long ago. Very little is known about coal-mining costs, but evidence pieced together from a limited amount of published data and interviews with officials of the coal and electrical power industries suggests that in 1973 eastern coal for electric utilities costs thirty to thirty-five cents a million BTUs to mine by underground methods. Concrete evidence is provided by reviewing the delivered-fuel prices reported by electric utilities and making rough adjustments for transportation costs. The impression is confirmed by information secured in interviews.

In 1969 the prevailing cost was eighteen cents a million BTUs. The sharp increases are the result of a variety of difficulties that have substantially raised wage rates and lowered output per worker. The essence of the problem is a reaction to the difficult, dangerous, and unhealthy working conditions in underground mining. Some of the change has been imposed administratively. A much stricter federal mine health and safety law was passed in 1969 in response to a particularly severe coal-mine disaster. However, general worker unrest, evident in official and wildcat strikes and greatly increased transfers among jobs, has also contributed to the change. The rise in wages, of course, was entirely the result of worker pressures. An influx of inexperienced workers has created additional difficulties.

These increases in labor costs per ton have been supplemented by rising costs of materials and construction. These cost increases will probably continue at least to 1980. So long as coal mining remains an unattractive occupation, its wage rates are likely to rise more rapidly than those in other occupations. To restore old levels of output per worker will be difficult. Even if inflation vanishes tomorrow, eastern coal for electric utilities will probably cost forty to fifty cents a million BTUs by 1980. Thereafter price rises may be more modest. Some enthusiasts believe that adequate research will ensure the development of new technologies capable of lowering coal mining costs by 1980. However, this belief is based on little more than faith. The sources of these new methods are vaguely sketched in general references to automated mining.

In contrast, western coal costs as little as fifteen cents a million BTUs and may only rise to twenty-five cents by 1980. Such cost increases would largely reflect extra expenses of a more rapid expansion and the need to use slightly less attractive resources.

The future of western coal will be affected by transportation costs, technological developments, public policies, the type of consumers, and many other factors. Because there are so many factors, it is impossible to determine the general effects of the development of western coal reserves. But an examination of effects in a few specific areas may be of some value.

The first is electric-utility fuel in Chicago, which represents a crucial market for western coal. It will be assumed that stack-gas scrubbers will be available for coal-fired plants, that strip mining in the West and nuclear power will not be banned, that western coal will still meet pollution regulations without special processing, and that oil-import policies continue to impose import fees rather than quotas. It will further be presumed that, whatever the delays, by 1980 uncontrolled burning of high-sulfur coal will be prohibited. It is prohibited already in the city of Chicago but could be burned in a plant located somewhere downstate.

The initial fuel-price assumptions are that western coal will cost sixty cents or more per million BTUs in Chicago, eastern coal fifty to sixty cents, and oil eighty cents. The first two figures are derived by combining the mining and transportation cost estimates. The oil figure is based on 1972 delivered costs of low-sulfur oil to the Middle West.

To complete the analysis, it is necessary to consider also the costs of different types of power plants and of sulfur-oxide control systems. An elaborate system of simplifying assumptions has been developed to ease the comparison of these different costs. The most convenient approach is to show the equivalent values per million BTUs of coal displaced. The difference between the total costs per kilowatt-hour of nuclear power and the costs excluding those for fuel and sulfur-oxide emission control of a coal-fired plant represents the maximum cost per kilowatt-hour that can be paid for coal and emission control if the use of coal is not to be more expensive than nuclear power. Multiplying this maximum allowable cost per million kilowatt-hours by the number of kilowatt-hours produced by a million BTUs of coal produces an allowable cost per million BTUs. Available data suggest that the allowable cost will lie in the forty- to fifty-cent range.

In dealing with coal-oil comparisons, it is preferable to indicate the allowable difference in fuel prices. Several cases can be distinguished. With old plants, the highest oil-conversion cost may be equivalent to increasing oil prices by five cents a million BTUs; however, this difference may be offset by saving the costs of installing sulfur-dioxide scrubbing facilities. In new plants, the advantage lies with oil. Avoiding the installation of coal-handling equipment is equivalent to an eight- to sixteen-cent reduction in fuel prices. This advantage may be added to the coal prices required to remain competitive with nuclear power, which will cost roughly fifty to seventy cents.

The cost of sulfur-dioxide scrubbing facilities is subject to considerable uncertainty. Figures produced by United States government agencies suggest that costs would run as low as ten to twenty cents per million BTUs for scrubbers installed in new plants and twenty to thirty cents for retrofitting old plants.[3] However, other estimates run much higher. A study by the Commonwealth Edison Company of Chicago points out that the low-cost estimates of government sources cover only the scrubber itself and omit the additional construction required, the cost of waste-disposal, and interest costs during construction. When these costs are included, scrubbing costs in a new plant lie in the

[3] U. S., Sulfur Oxide Control Technology Assessment Panel (SOCTAP), *Final Report on Projected Utilization of Stack Gas Cleaning Systems by Steam Electric Plants*, 1973.

range of forty-five to fifty-eight cents. Commonwealth Edison believes that this method further understates the costs by assuming higher operating rates than can be sustained. It suggests a range of fifty to sixty-three cents. During the early life of a new plant, the most probable figures are between fifty-four and fifty-seven cents. Retrofitting scrubbers costs seventy-five to eighty-five cents.

Given the assumptions stated for Chicago, the choice for existing plants would be western coal. The sixty-cent cost would be less than the fifty- to sixty-cent delivered cost of Illinois coal and the scrubber cost of at least twenty cents. Western coal is less than the eighty-five-cent cost of oil and conversion. However, it is conceivable that even in Chicago, western coal will cost more than eighty-five cents because of a rise in rail rates and limits on the amount of strip mining allowed. Moreover, the cost advantage would decline, perhaps sharply, as western coal moved further east. Should oil be cheaper than western coal, eastern coal could survive only if the cost of scrubbers were as low as the United States government contends. When it becomes possible to build new fossil-fuel plants, oil will have an effective price of around seventy cents because of the savings in capital cost. Again it would appear that if oil costs stay in the specified range, oil would certainly be cheaper for new plants than fifty- to sixty-cent eastern coal with scrubber costs of twenty cents or more. However, unless western coal were significantly more expensive than initially assumed, it would be competitive in the Midwest. Of course, this position could be undermined if the practice being adopted in western states of requiring scrubbers with western coal were also imposed east of the Mississippi.

However, nuclear power would be the preferable alternative when compared to those already noted. Forty- to fifty-cent coal or sixty- to seventy-cent oil may be hard to obtain. The introduction of a combined cycle using low-BTU gas from coal or oil appears to improve coal and oil's position considerably but until the requisite technologies are employed the improvement may not be sufficient to stave off the nuclear threat.

Moreover, oil could prove the actual beneficiary of the combined cycle. If published figures on low-BTU gas are updated to remove the credits for unlikely sulfur sales and to account for a doubling of capital costs, coal gas would have a cost per million BTUs of 1.15 times the cost of a million BTUs of coal plus thirty-one cents. The corresponding oil gas figures are 1.1 times the oil price plus twenty-two cents. Moreover, the gasification costs include desulfurization. When the oil itself is desulfurized, about 17 cents is added to the cost of oil. Thus, instead of competing with eighty-cent oil, coal would compete with sixty-three-cent oil. Coal prices would have to be below fifty-two cents a million BTUs to be competitive.[4]

Given both the considerable uncertainties about actual cost developments and the possibility that one reaction to such developments would be a change in policies, these calculations should be taken only as indicators of future developments. No one familiar with the issues would argue that actual prices will closely correspond to the estimates presented here.

It can be argued, however, that it is very important to develop policies to resist rising oil prices. The doubts that oil can compete against either coal or nuclear power would vanish overnight if oil were priced closer to its true costs. In the case of coal and oil, the introduction of additional environmental considerations greatly favors oil. Its environmental impacts are both milder and easier to control than those of coal mining. It is not clear whether the environmental problems of oil are truly worse than those of nuclear power, mainly because nuclear dangers have not been determined.

Moreover, it appears likely that oil is going to prove superior at least to eastern coal as the fuel in conventional fossil-fired plants. Because the Commonwealth Edison estimates of scrubbing costs are probably more realistic than those of the United States government, oil costs could rise considerably and still be competitive. Western coal, on the other hand, may indeed penetrate markets east of Chicago so long as transportation costs can be kept down and strip mining is not severely limited.

Nuclear power is likely to become the main source of electricity generation if coal prices rise as predicted here and oil prices do not fall sharply. In the near term, oil or western coal and then nuclear power will displace eastern coals as the basic power-plant fuels. A comeback is possible only if the combined cycle and coal gasification prove successful and mining-cost rises are limited.

Public policy will be able to play only a limited role in preventing the shift away from eastern coal. It is conceivable that coal might not be competitive with nuclear power even if sulfur-oxide control rules were completely eliminated. However, it is quite clear that temporary suspension of the rules might slow down the decline in the consumption of eastern coal by the electrical power industry. The wisdom of the

[4] F. L. Robson et al., *Technological and Economic Feasibility of Advanced Power Cycles and Methods of Producing Nonpolluting Fuels for Utility Power Stations*, PB 198 392 (Springfield, Va.: National Technical Information Service, 1970).

suspension clearly can be disputed, but the issues are far from clear-cut. The regulation of sulfur emissions appears to be designed to solve problems in heavily polluted regions and may be excessive in many parts of the country. Thus in the coal industry more turmoil than necessary may be created by pollution regulations.

Even less is known about gasification and liquefaction than about electrical power. All that is really known is that when using strippable western coal, high-BTU coal gas is likely to cost at least $1.25 a million BTUs. Although such gas is more difficult to manufacture than that used by power plants, this upgrading is essential in providing a satisfactory substitute for natural gas. Oil from similar coals might cost as little as $6 or as much as $8.

One critical problem is that adequate supplies of cheaply strippable coals may not be available for massive coal-synthesis industries. The known cheaply strippable coals only meet a modest portion of America's oil and gas needs. Since further exploration may not unearth the additional reserves needed, the economical production of synthetic oil and gas from coal is subject to exactly the same uncertainties of other energy sources. The situation is aggravated by the growing resistance to strip mining in the West. The vast amount of mining required for producing all oil and gas supplies from coal may not be tolerated.

The question remains of whether western coal will be the cheapest source of oil and gas. For example, oil shale and tar sand might prove preferable as a source of oil. Consumers might shift to oil rather than use high-cost gas.

Is the Breeder an Inherently Inferior Alternative?

An economic argument against the present Atomic Energy Commission (AEC) case for the breeder has deservedly attracted considerable attention. The breeder appears to be inferior to conventional nuclear reactors.

Briefly, the special characteristics of the breeder imply that it has lower fuel costs than a conventional reactor. However, these lower fuel costs will be obtained by investing in an expensive development program that will lead to commercial plants with higher costs than ordinary nuclear plants. The official AEC justification of the breeder employs several dubious assumptions.[5] It not only underestimates the

---

[5] U. S. Atomic Energy Commission, *Updated 1970 Cost Benefit Analysis of the U. S. Breeder Reactor Program*, Wash. 1184 (Washington, D.C., Government Printing Office, 1972).

domestic supply of uranium, but assumes a continued ban on uranium imports. It can be argued that the only purpose of such a ban is to justify the breeder; it is hard to imagine what risks the United States runs in buying Australian uranium. In addition, the benefits of the breeder program are exaggerated by use of a more optimistic time-table and lower cost of capital than seems plausible.

Ironically, the only way to save the argument involves totally dis-crediting AEC's analysis. It could be argued that costs will be so much lower than expected that the breeder will be viable even if uranium prices do not rise. Some nuclear engineers contend that this expecta-tion is quite reasonable. While this hypothesis is questionable, its acceptance would indicate that every major AEC premise was incor-rect. One would then wonder why such large sums of money were devoted to such a project and be skeptical that the revised argument had any greater plausibility.

Conclusion

In the next generation only nuclear power can be a reasonable alterna-tive to conventional sources of oil and gas. Data indicate only that ex-pensive efforts to promote cheaper production and utilization of alter-natives for the relatively near future may prove futile. The uncertain-ties are so great that research support should be designed so that it might be terminated if failure should occur.

The greater danger of stressing alternatives is that it will lead to abandonment of efforts to rationalize the world oil market and to un-necessary protectionist measures to save the United States from the consequence of unwise commitment to inferior fuel sources. This is no mere theoretical conjecture. The history of energy policy is full of ex-amples of precisely this error. The United States Oil Import Program was originally designed to protect high-cost domestic oil producers and was quite ineffective in preventing the real threat of a cartel. Europe has spent billions of dollars to protect an ill-advised decision to preserve its hopelessly high-cost coal industry.

On an instinctive basis, it can be suggested that the more quickly developed alternatives such as oil shale and coal synthesis are too un-attractive to waste money on. It may be better to skip them and con-centrate upon ensuring that fusion or a similar twenty-first century option is made economical as rapidly as possible. Public policies should recognize the drawbacks of shifting from conventional sources of oil and gas.

Waste-heat Utilization

BARRY L. NICHOLS

Waste-heat utilization as a means of conserving energy is not a new concept, but the incentive to implement programs of waste-heat utilization is new. All energy use results in the production of heat. To capture and use this heat before it is lost from the planet would permit a more effective use of the nation's energy resources. Waste heat can be substituted for fossil fuels in agriculture and aquaculture within closed buildings. Food production can be increased by optimizing growth conditions for plants and animals. These and many other uses of waste heat would not solve all the energy problems of the next decade, but they would conserve energy and reduce the imbalance between demand and supply.

Few will dispute the importance of abundant energy in the nation's development. Through its massive energy capacity, the United States, with a population of 200 million, marshals the equivalent manpower effort of 100 billion people. Abundant energy has made possible the production of goods and materials and a standard of living that would have otherwise been unattainable. But, through inescapable by-products, the rapid growth of energy increasingly affects American society in unforeseen ways. One of the inescapable by-products of energy-conversion processes is heat of a low temperature, itself another form of energy. This heat is sometimes difficult to discard, and, because of its low temperature, is always difficult to use. The problem is how some of this waste heat might be substituted for primary-energy consumption as one way of conserving energy.

Sources of Waste Heat

The principal energy-conversion processes in this country include those used for transportation, industry, residential and commercial needs, and generation of electricity. Of these, the last is the fastest-growing conversion process. While the population doubles approximately every fifty years, the consumption of electrical energy doubles in less than ten years.

A basic principle of nature, enunciated as the second law of thermodynamics, is that ultimately all energy, whether mechanical, electrical, nuclear, or chemical, appears as heat when converted. Electricity is generated with efficiencies ranging from about 33 to 40 percent, depending principally on the type of power plant. Unfortunately, efficiency has been increasing slowly over the past two decades, so that the waste heat produced has been increasing almost linearly with the growth in electricity generation.

Compared with other energy-conversion processes, the generation of electrical energy is an efficient one. The internal-combustion engine, for example, converts chemical fuel to mechanical energy with an efficiency of less than 30 percent. Electricity generation, however, is carried out in large plants, where vast amounts of waste heat are produced and normally dissipated at the point of generation.

"Waste" heat designates energy that is so degraded in temperature that its uses are limited, and usually it is considered practical only to discharge it directly to the environment. Such energy appears in the large quantities of cooling water used for condensing steam discharged from the turbine in steam-electric power plants. Typical outlet temperatures for such cooling water are in the range 60° to 95° F., depending on the ambient water temperature, the quantity of water circulated, and other factors. For those power plants that have evaporative cooling towers, the outlet water temperature would be increased by 15° or 20° F. to the range of 75° to 115° F., while for dry cooling towers it would be increased by 20° to 40° F. to the range of 80° to 135° F.

Over the past two decades, the capacity of electrical power plants has increased significantly. In 1950 the size of the average generating unit placed in operation was only 48 MW(e), but by 1970 the size of the average nuclear-plant unit scheduled for operation was over 700 MW(e). The construction of multiple units at a given site further increases the difficult task of handling the large but local sources of waste heat.

In 1970 the electrical generating capacity from steam electric plants in the United States was approximately 265,000 megawatts, and in 1973 it is estimated at over 300,000 megawatts. These plants produce about 8×10^{15} BTU/yr of waste heat. By 1970 electrical-energy generation accounted for approximately 24 percent of total energy consumption in the United States. By 1980 it is expected to account for approximately 30 percent. About 15×10^{15} BTUs/yr (or the equivalent of 500,000 MW) will be discharged to the environment continuously as waste heat. Over the next several decades a significant number of light-water reactors, which have lower efficiency, will be introduced. In recent years the efficiencies for new fossil-fired plants have been leveling off. In addition, the greater use of environmentally associated equipment, such as cooling towers, decreases the net plant-heat rates. Therefore, it is unlikely that there will be any substantial improvement in the average heat rate and therefore inefficiency over that for presently operating generating plants.

If even 5 percent of this waste heat can be effectively utilized for productive purposes, and if these productive purposes can replace energy that would otherwise be required to accomplish similar tasks, then the energy annually saved would be equivalent to 7.5×10^{14} BTUs, or a steady generating capacity of 25,000 MW of equivalent heat. In the use of natural gas, for example, this represents a heat equivalent of approximately 750 billion cubic feet of gas or between 3 and 4 percent of the 1970 rate of consumption in the United States. If it were possible to use 10 percent of the heat rejected from the generating stations to be built over the next thirty years, the net effect would be to use more waste energy than the equivalent in electrical energy generated today, approximately 300,000 MW(e).

Over the next two decades, present fuel sources such as oil and natural gas will be in increasingly short supply. It is anticipated that the demand for oil and natural gas will increase, their availability will decrease, and their price will increase. Many of the potential areas in which waste heat could be used, such as for greenhouses, are currently supported preponderantly by oil and natural gas. Thus, the substitution of waste-heat sources for oil and gas represents not only a potentially less costly fuel, but also a means for reducing, even if in a small way, the demands on shrinking fuel resources.

For these reasons there is significant interest in methods for utilizing rejected heat. The use of waste heat might also afford the opportunity to reduce the adverse environmental impacts of waste-heat discharge, to reduce the cost of handling thermal discharges, and to improve energy utilization. In addition to the heat from large, central-station power plants, the discharge heat from package energy-conversion units, which supply space heat or electrical energy to smaller load centers such as shopping centers or other commercial buildings, can also serve as sources for usable waste heat. In fact, the small energy-centers often provide a closer match of available heat supply to the amount of waste heat that can reasonably be used at a single site.

While elimination of thermal pollution may not be possible, the impact from thermal discharges may be reduced. Incentives do exist to use heat now wasted.

Uses of Waste Heat in Closed-structure Agriculture

Agricultural uses, such as environmental control of animal shelters and greenhouses, offer a way to use thermal discharges from power plants and other industrial processes by means of once-through or closed-cycle cooling systems. Power plants with cooling towers are normally designed so that the temperature of the condenser discharge water is between 80° and 100° F. Plants with once-through cooling normally produce somewhat lower temperatures but can provide the bulk of the heat needed for most of the year. These temperatures are high enough to provide optimal thermal environments for many plants and animals. Maintaining animal shelters at the proper temperatures can increase growth rates and feed efficiencies, particularly for smaller animals such as poultry and swine. Greenhouse production of both flowers and vegetables is critically dependent on artificial heating and cooling. The use of cooling water from condensers for heating can significantly reduce fuel costs and serve as a substitute for heat produced by burning oil or gas. Thus, some agricultural operations may truly be considered potential waste-heat users.

In spite of these obvious benefits to agriculture, the current level of production from greenhouses is such that only a small fraction of waste heat discharged from power plants can be profitably used. The projected growth patterns suggest that this picture will not improve in the future. In addition, the use of waste heat is strongly dependent on the geography, climate, and season. Nevertheless, agricultural uses of waste heat are sufficiently attractive, under certain conditions, to warrant serious consideration. While these uses will not solve the problem of thermal pollution, they can reduce the impact of thermal effluents on local ecology, conserve energy resources, and save money for both the supplier of waste heat and the agricultural operator.

There are no large agricultural operations in this country using low-level heat from power plants, but some experimental work could lead to large-scale use in the future. Research at the University of Arizona, the University of Sonora in Mexico, and at Oak Ridge National Laboratory suggests that using waste heat for agricultural climate control is both feasible and economically attractive. The University of Arizona and the University of Sonora are using waste heat from diesel generators to provide heat for greenhouses at Puerto Penasco, Sonora, Mexico, where crops have been grown at near 100 percent relative humidity. One unique feature of the Arizona facility is the ability to conserve water through collection of condensed water on the plastic roof of the greenhouse. This power system provides water and food. The success of the Mexican venture has led to a project in the sheikdom of Abu Dhabi for construction of a similar facility on the island of Sa'Diyat.

A study at the Oak Ridge National Laboratory indicates that there are several advantages in using greenhouses to cool condenser water, including the return of cooler water from greenhouses as compared with cooling towers. Recently work has determined the actual operating performance of such systems for use in heating and cooling of greenhouses. Waste heat contained in the water from an air-conditioning system was used for year-round temperature control in a small experimental greenhouse. The Tennessee Valley Authority is constructing a test facility to use the "Oak Ridge system" of heating and cooling in a joint TVA-AEC program.

One system envisioned for heating and cooling both plant and animal shelters involves the use of a conventional pad and fan system, but also includes a finned-tube heating coil. Pad and fan systems are currently used in many greenhouses and in some poultry and swine operations for cooling purposes. The pad and finned-tube system is used in the Oak Ridge experimental greenhouse.

The pads are typically filled with a fibrous material such as aspen. Condenser cooling water flows onto the pads from a trough at the top and passes down along the fibers. Air flows horizontally across the pads and is heated or cooled depending on the relative ambient air and water conditions. The cooled water is collected at the bottom of the pads and pumped back to the condensers.

Warm water from the condenser may also be pumped through the finned-tube coils, located downstream from the pads. The air, coming from the pads, is heated and dried by the transfer of radiant heat from the fin coil. By varying the relative amounts of water pumped through the pads and the coils and by changing the flow rate of the air, the temperature and humidity of the air entering the enclosure can be adjusted over wide ranges. This system can be used for both summer cooling and winter heating. The heated or cooled air passes through the house and out the other end through exhaust fans. Automatically controlled louvers permit recirculation of the air under conditions such as extreme cold.

With this system the environment within the enclosure can be maintained near the optimum. Simultaneously, the condenser water is cooled to near the wet-bulb temperature. Thus, the enclosure serves as a horizontal cooling tower. This system also involves the flow of warm water to the greenhouse, where the water is cooled and sent back to the source or discharged to surface waters.

It is estimated that if waste heat from existing sources were available to heat existing greenhouses, only a small percentage of the waste heat could be used. Therefore, the primary incentives must be economic rather than the hope of solving thermal-discharge problems.

As with cooling towers, the impact on the environment of the consumption of water must be considered. Water required in heating and cooling systems is partially consumed by evaporation, and some is required for the growth of plants. The degree of the impact on the environment from water consumption has yet to be determined, but it appears that the loss would certainly be less for greenhouses, if the water condensing on the greenhouse surface was collected and returned, than for cooling towers. During recirculation in winter, most of the water could be recovered from condensation.

The use of waste heat to provide optimal temperature conditions can reduce fuel bills, conserve energy, and increase feed efficiency and growth rate for both hogs and broilers. A pad and fan system in conjunction with a finned-tube coil, similar to that described above, can provide both winter heating and summer cooling.

Thus, the use of waste heat for warming animal shelters could save poultry and swine operators many dollars in fuel costs based on current fuel-consumption figures. However, current practice may not maintain optimal temperatures; hence the potential savings may be significantly higher if the additional savings that would result from improved feed efficiency and increased growth rates are considered. Feed accounts for over 60 percent of the total cost in both broiler and swine operations. Increasing the ambient temperature from 60° to 70° F. increases the feed efficiency for broilers by 0.05 pound-feed/pound-gain. Similarly, increasing the air temperatures from 60° to 65° F. for swine reduces the total feed consumed by 20 pound/hog. Thus, even slight changes in ambient temperatures can reduce feed costs.

Use of Waste Heat in Aquaculture

Aquaculture has been practiced for centuries in the Orient, particularly in the tropical and subtropical areas where farmers raise fish in flooded rice fields to provide a protein supplement to their basic grain diet. Yet, the practice is also a new technology. A few fish species have been intensively cultivated in controlled environments, and yields of these species have been enhanced by the degree of management exercised over the operation.

Basic data on fish growth indicate the potential benefits of temperature control. For example, shrimp growth is increased by 80 percent when water is maintained at 80° F. instead of 70° F., and catfish grow three times faster at 83° F. than at 76° F. Growth of both aquatic species benefits appreciably more from temperature control than animals such as broilers, cows, and swine. All optimum temperatures for growth are within the range of temperature of power-plant effluents. Thus, heated discharge water might be used to increase the growth and yield of animal protein.

Heated discharge water from steam-power plants represents a large thermal-energy source for maintaining the temperature of a culture medium in a range that is optimal for the growth of aquatic species. Electricity is available at low cost for pumping power that permits greater environmental control over the water system. Thermal aquaculture offers the potential of producing high-quality aquatic foods continuously and the possibility of decreasing the present variation in supply that results from the seasonality of such produce.

Temperature control alone, however, is not sufficient for optimum production of aquatic species. Dissolved oxygen content, biological oxygen demand of the culture system, fish waste control, and nutritional adequacy of the diet are some of the other important variables that influence yield.

Power-plant coolant water has only recently been used for aquaculture. A commerical operation, the Long Island Oyster Farms of Northport, Long Island, utilizes the thermal effluent of the Long Island Lighting Company Northport Plant for the early stages of oyster culture. Normal growing periods of four years have been reduced to two and one-half years by selective breeding, spawning, larvae growing, and "seeding" in the hatchery heated by effluents. These practices avoid reliance on variable natural conditions and permit accelerated growth in the plant thermal-discharge water for a period of about four to six months, when the water would otherwise be too cold for maximum growth. Oyster culture is completed for market in the cold waters at the eastern end of Long Island Sound. The product is harvested, processed quickly to the frozen state, and marketed for $15 to $20 a bushel (1971 price), the upper end of the wholesale price range. About 20 percent of the oysters set in the hatchery result in a harvested product.

Catfish have been cultured in cages set into the thermal discharge canal of a fossil-fueled plant of the Texas Electric Service Company of Lake Colorado City, Texas. During the winter of 1969–70, growth rates were equivalent to 100 tons per acre a year. This is comparable to the yields of rainbow-trout culture in flowing water. The Texas operation is now on a commercial basis.

A catfish experiment is being conducted by Cal-Maine Industries and the Tennessee Valley Authority at the TVA steamplant in Gallatin, Tennessee. Heated discharge water from the plant is circulated through ten concrete channels, each four feet wide, four feet deep, and fifty feet long. Algae formation is minimized by covering the channels and preventing photosynthesis. Inlet and ambient water are blended for temperature control. Nutritionally balanced feed in pellets is fed to catfish in culture, and an experimental aeration system is under test.

In the agribusiness industry, large feed-production and animal-processing companies are considering the use of waste heat for fish cultivation. The Florica Power Corporation of St. Petersburg, Florida, is participating in joint research with the Ralston Purina Company to develop a satisfactory technique for culturing shrimp at the utility's Crystal River site.

Smaller companies like International Shellfish Enterprises are developing methods for oyster culture in the thermal discharge canal of Pacific Gas and Electric's plant at Humboldt Bay, California. Marifarms of Panama City, Florida, is using the warm water from local power plants to maintain pond temperatures in winter so that the mortality rate of shrimp in culture is minimized. An experimental lobster culture using warm water is being considered by a few institutions, including the San Diego Gas and Electric Company, the Mariculture Research Corporation, and the Department of Sea and Shore Fisheries of the state of Maine.

The Japanese have led the way in demonstrating the benefits of waste-heat utilization for aquaculture. Shrimp, eel, yellowtail, sea bream, ayn, and whitefish are being cultured. Culture experiments started at the Sendai Power Plant in 1964. Five other demonstration programs have been established at generating stations powered by fossil fuels. In a pond culture at a power plant in Matsuyama, shrimp are cultured in thermal effluents which are blended with ambient water in order to maintain constant temperature. During the summer, shrimp grew to one and one-fifth times the weight of shrimp in natural summer-water temperatures, while winter growth was seven times that in ambient-temperature water. Survival rates were about 50 percent in the summer and in excess of 30 percent in the winter. In flowing water, yellowtail cultured at a constant temperature from October to June grew to a weight of one and one-half times that of fish cultured in natural water. No mortality or parasite problems were encountered. At the Tokai-Mura nuclear power station near Tokyo, a multispecies demonstration program of thermal aquaculture has been approved by the Japanese government. The five-year program costing $575,000 is to develop a facility consisting of thirty-five concrete channels of various sizes to demonstrate flowing-water cultures.

The English have had a small development program since 1966 for the culture of flatfish species, plaice and sole, at their nuclear plant in Hunterson, Scotland. The problem of free-chlorine toxicity was avoided by the use of a continuous chlorination treatment of coolant water instead of the conventional batch treatment. No radioactivity is allowed to be diluted into the coolant water stream that is used for aquaculture. Although culture of the flatfish species has been demonstrated, present methods of culture are not economically attractive because of the low efficiency of food conversion. Low-value fish are used as feed; a suitable low-cost, formulated food has not been developed. Flatfish are cultured in flow-through ponds near the shoreline. Since the system is not isolated from the sea, predators and disease are of concern.

Technological Problems and Development

Large-scale use of waste heat for aquaculture would probably not be considered until demonstration projects at existing sites indicate economic viability. Engineering design and evaluation of intensive aquaculture systems are needed. Applied research and developmental work would be necessary to complement engineering tests. For a given species, mass-culture techniques can be quite different from the techniques used in laboratory experiments. Flow rates for channel culture must be optimized so that energy of physical activity is minimized and food-energy conversion into flesh is maximized. Aeration systems should be evaluated. Handling devices for transferring and harvesting in a flowing system need to be considered, and systems to treat waste need to be designed.

Selective breeding should be conducted to produce species particularly amenable to intensive culture, and fish culturists must be able to furnish fingerlings the year round in order to have truly continuous culture. For example, finfish could be grown in channels, fish waste could be converted to algae, and intensive oyster culture fed on this algae.

Combined agriculture-aquaculture systems might be considered, particularly in the summertime when effluent temperatures may be too warm for fish cultures. Greenhouses might be used as cooling towers to extract heat from thermal effluents, and the discharge from greenhouses could be used for fish culture. This integrated system might permit the maximum use of waste heat for food production and simultaneously incorporate aquaculture into a closed, recirculating system instead of a once-through cooling system.

The economical utilization of large quantities of energy at low temperature is a task that would have already been solved if it were not very difficult. The incentives, however, for the perfection of waste-heat utilization have increased and seem likely to continue to increase in the future. Future energy demands are expected to increase the production of waste heat. Rejection of this waste heat by conventional means will result in the loss of water because of increased evaporation, and this may present increasingly grave problems for the environment. Use of this heat in agriculture and aquaculture serves as a means of reducing fuel consumption, increasing the effectiveness of fuel use and the yield of crops, thus reducing the amount of land needed for agriculture, and, for some applications such as greenhouses, reducing water consumption.

The potential market for thermal energy may represent an additional, significant source of income for utilities while providing the user with low-cost heat. Studies for a small power plant, which assumed a market for all the waste heat generated, indicated that the value of the waste heat produced was over $1 million a year. If this much heat could be sold, it would represent a significant proportion of the revenues from electrical sales. Even if only a fraction of the waste heat can be marketed, it may help offset added operating cost of installing equipment related to environment protection, such as cooling towers.

The methods described here cannot reduce the total thermal energy dissipated from power plants and other energy sources. However, these processes can provide economical ways in which waste heat can be substituted for the heat produced by other energy sources. In this way energy can be conserved and thermal discharges reduced on a national and global scale. It can also permit the introduction of heat into the biosphere in a more acceptable manner and thereby reduce other consequences harmful to the environment.

The sale of waste heat by the electric utilities may present special problems. Can the added sales be used to increase the profit to the utility, or must they be used to reduce the rates paid by the customer? This question is one example of the institutional problems to be solved. The question of who owns the water after it leaves the power plant and goes to the waste-heat user must be considered for cases in which the water is discharged into the environment or in areas where the appropriation doctrine of water-use prevails. Responsibility for the effects of thermal pollution must still be accepted by either the producer of the waste heat or the user.

Discharges of waste do not differ greatly, whether or not a facility uses waste heat as a source of energy. The possibility of discharging heated water is the only difference. How waste heat can be used without conflicting with water-quality standards for thermal discharges has yet to be determined. Some indication of how it will be addressed can be found, however, in the Federal Water Pollution Control Act Amendments of 1972. Section 318 of the act allows the discharge of specific pollutants from aquaculture facilities under federal or state supervision.

The use of waste heat to conserve energy will be truly viable only when problems such as these have been solved satisfactorily. The impending energy shortage will no doubt prove to be an incentive to solve these problems and should encourage the use of waste heat on an expanding scale in the next decade.

Exploitation of the Continental Margin

HERBERT D. DRECHSLER

The seabed location of oil reserves is perhaps the most important new source of oil and gas in the United States. In 1954 the seabed provided about 2 percent of the nation's oil supply; by 1973 it was providing 17 percent of the total United States production. Estimates for 1995 put offshore production at over 30 percent of the total United States petroleum output.

The transition of the United States petroleum industry from onshore to offshore production is the result of two factors. The first is a higher degree of success in offshore exploration. From 1948 through 1967 in the United States, 3,800 exploratory wells were drilled for each giant field discovered; however, in Louisiana only 155 offshore exploratory wells were drilled for each giant field discovered. The second factor is that larger quantities of oil are discovered offshore than on land. In the Gulf of Mexico, from 1948 through 1967, wells produced about four times as much oil as onshore wells in terms of footage drilled. As a result of these factors, production offshore is about as profitable as on land even though the cost of an offshore oil well is about six to nine times that of an onshore well.

The term *offshore* may be defined according to its geological meaning (the physical area) or according to its political concept (the human-defined boundaries of such areas). The two concepts usually conflict when the national interest is involved.

Between the continental land masses lie four distinct ocean regions. The first is the continental shelf, which starts at the coastline and dips gradually downward. Eventually, the slope becomes increasingly steep at an observable break, which marks the margin of the continental shelf. The continental shelf has a worldwide average width of forty-eight miles but varies from zero to over nine hundred miles.

The second region, the continental slope, is seaward of the continental shelf. The boundaries are the edge of the continental shelf and the point where the slope begins to flatten as the bottom approaches the abyssal depths. This seaward change of slope is discernible over many miles but not easily observable at any particular point. The major identifying feature of the continental slope is the change of slope from increasing steepness on one side to decreasing steepness on the other side. The water depth of the continental slope ranges, in general, from 4,000 to 12,000 feet.

Seaward of the continental slope is the third region, the continental rise. This region is bounded by the continental slope and the deep ocean floor—the fourth region—on the seaward side. The continental rise is typified by a flattening of the continental slope.The three areas considered together, continental shelf, continental slope, and continental rise, are called the continental margin.

Off much of the east coast of North and South America and the west coast of Africa, the continental rise is quite wide. Its width seems to be related to the stability of the continents. Where the continental land masses are moving, the rise is narrow because sediments from the rise and slope have been carried to the deep ocean floors.

Underlying the continental margin is rock covered by the sediments in which petroleum resources form. The upper slope sediments usually contain more organic material than the shelf, but there is substantial geological evidence that conditions at certain times in the past have favored the accumulation of petroleum under the lower slope and rise. The locations where petroleum resources may be found, in order of importance, are the continental shelf, the continental slope, shallow seas and small ocean basins such as the Gulf of Mexico, and the continental rise.

By 1985, petroleum is expected to provide almost 45 percent of the energy required to meet the United States's needs. Because much of this oil will be produced in other countries, the international community is deeply involved in the United States energy problem and also in seabed exploration.

Many types of countries are concerned about petroleum production. Some basic categories are coastal state, land-locked states, industrialized states, nonindustrialized states, oil-producing states, and states that do not produce oil. In 1969, petroleum was produced in over seventy countries. Currently, 135 countries have some offshore activity, primarily for oil and gas.

The specific interests of these states fit into several categories. The first is the desire to regulate production and prices of crude petroleum and its products. Regulation of output is necessary for the stability of prices. The paradox of economics—the desire for entrepreneurial monopoly opposed to the deterministic force of competition—greatly affects the income of states through price fluctuations.

The desire for the ability to control exploration and production of oil from seabed resources is the second category of interest. Major industrial countries have the capital and managerial know-how necessary to explore for oil, but most of the other states do not. These poorer states may want to enter the oil-production and marketing business rather than remain subservient to the more wealthy states.

A country's power and ability to control its environment is the final category. Broader than the recent development toward pollution control, this concern includes the social, political, and economic environment. Seabed production of oil may substantially affect the total environment of nations, as oil production has in the past. Today countries want the knowledge and power to control their own national systems.

Coastal States and Land-locked States

The possible conflicts and trade-offs between coastal states and land-locked states is in a sense an argument between the haves and the have-nots. The coastal states have some degree of continental margin. From their point of view, the question is how much of a margin they have. Traditionally, countries have claimed a three-mile boundary offshore, suggested perhaps by the distance a cannon ball could be accurately shot. More recently, some countries have extended their boundaries as much as 200 miles offshore. However, most countries, including the United States, recognize a twelve-mile boundary.

Countries have often regarded territorial boundaries as a means of national defense and of protecting resources. Today they recognize that the continental margin, slope, and rise also have possible value as sources of oil. Thus the battle for offshore boundaries has begun.

In 1945, President Truman declared that the United States has exclusive sovereignty over the resources of the seabed of the continental shelf. The shelf was defined as the seabed up to a water depth of roughly 200 meters. In effect this declaration was a unilateral extension of sovereignty with alternative horizontal limits.

The seabed boundaries can be defined by geology, depth, and distance. The traditional boundaries have been distance, a 3-mile, 12-mile, or 200-mile limit. Truman developed a newer boundary concept that tried to encompass both a depth and geological boundary by combining the continental-shelf and depth criteria. This doctrine was not completely accepted, because some nations, such as Chile and Peru, have no continental shelf and because inland seas without horizontal boundaries, such as the Mediterranean, become battlegrounds.

The Truman doctrine did develop a boundary at the 200-meter line that was quickly accepted by some nations. In 1958 the Geneva Convention on the Continental Shelf added the exploitability clause to the Truman doctrine. It voided the 200-meter criteria and extended the horizontal boundary seaward to the depth where the superadjacent water allowed exploitation of the natural resources of the seabed and soil. This type of boundary developed a new concept of territory based on technology. Twenty-five years ago the oil companies were producing in fifteen meters of water. Today wells produce from 150 meters of water. Since oil will be located in all parts of the continental margin, the boundary is flexible and dependent on the ingenuity of petroleum engineers. In effect the boundary has been extended to the limits of the continental margin.

Oil-industry engineers have recognized a problem that the politicians ignored or avoided. In 1969 the National Petroleum Council, the industry lobby, stated: "The United States, in common with all coastal nations, now has exclusive jurisdiction over the natural resources of the submerged continental land mass seaward to where the submerged portion of that mass meets the abyssal ocean floor and that it should declare its rights accordingly." Using the existing rules of the political game, the oil industry clearly pointed out that eventually the entire continental margin would be subject to oil exploitation.

In August 1970, the government presented the Draft Convention on the International Seabed to the United Nations. This proposal brought back the coastal-states boundary to the 200-meter line and created the International Trusteeship Zone between the 200-meter line and the deep ocean floor. The zone would be administered under joint coastal-state and international jurisdiction. International jurisdiction, which is not well defined, is the essence of the problem. Petroleum resources may extend to the limit of the continental margins. The questions that are still unanswered are: Who owns these petroleum resources? Who shall gain revenue from them? Who shall administer the production from them? The political arguments surrounding these questions are related to the classic economic problems of the profit motive, economic growth, absolute economic growth, and changing technological systems.

The land-locked states also wish to know whether they will receive income, how they can enter the oil business, and how they will control production. Limits of space preclude discussion of the legitimate concerns relating to the energy needs of land-locked states, such as Austria and Switzerland, and constrain this discussion strictly to territorial problems.

Since the land-locked states have no seacoasts, they view themselves as an underprivileged group. But this group of twenty-nine land-locked states have two primary levers that enable them to exert pressure and protect their interest. The first is the principle, accepted by the United Nations in 1967, that the seabed should be utilized for peaceful purposes and is the "common heritage of mankind." One aspect of the common-heritage principle is the acceptance of patrimonial wealth. That is, governmental or administrative revenue received from seabed exploitation should be shared by all nations. The 1967 declaration stated that the net financial benefits should be used primarily to promote the development of the poorer states. It should be noted that poverty is a relative concept and is measured by gross national product per capita, which is an imprecise and inaccurate tool.

However, all twenty-nine states would have a voice in the distribution of net revenues. If the quantity of money received by the poorer land-locked nations were great, all land-locked nations would probably want some income. The method of dividing the income among the land-locked nations is a question that requires further negotiation.

Each of the twenty-nine countries has a vote in the United Nations. These votes can be moved in particular directions depending on the persuasive powers of the other voting members. In efforts to gain a favorable seabed treaty, some nations such as the United States could trade off other relevant interests to some or all land-locked nations. The vote of the land-locked nations becomes a valuable negotiating item in respect to the interests that may have little to do with the seabed treaty.

One must keep in mind that the twenty-nine land-locked nations are about one-third of the United Nations membership and as such are a significant voting bloc. While one should not necessarily assume that these nations would vote as a group, in view of the possible petroleum riches believed to lie in the seabed, there are significant reasons for a bloc vote.

Industrialized States and Nonindustrialized States

The industrialized states have a substantial growth rate ranging from 4 to 6 percent a year and a high absolute need for energy. In nonindustrialized countries, the growth rate for energy may be higher than in industrial countries; however, the quantity that they use is comparatively small. For the next ten or twenty years the industrialized nations will have a greater need for seabed supplies of oil while the less industrialized nations will not be as dependent upon them. Perhaps by the year 2000 the cumulative effect of high growth rates in some of the less industrialized nations will increase the absolute needs for energy enough to make the seabed a more significant supply location.

Many nations have neither industry nor high growth rates for energy. For them, the seabed sources of oil have little value except as a potential source of revenue. To be sure, the natural desire for improved social and economic conditions has hastened a desire for growth. The seabed may provide the energy sources for these nations many years in the future. Their concern is that only high-cost supplies of oil will be available when they need them.

Because petroleum resources are generally produced in a sequential order from lower to higher cost, current exploration has shifted toward the seabed where the cost of finding and producing oil is less than on the land. People are influenced by Malthusian theories when they believe that natural resources over the relatively near term, say 100 years, are exhaustible and that all of the oil that existed on this planet will be consumed in the next 100 years. One might well argue, however, that oil reserves will not be consumed in that time and it does not matter if they are consumed.

Petroleum resources are not exhaustible, because they are renewed by a process called exploration. If the oil that is discovered, however, can only be produced at a cost higher than the prevailing price, it is left in the ground for future development when the low-cost resources are exhausted. Then the oil that will be produced is also relatively low-cost because all other available oil could be produced only at a higher cost. The oil produced at any time, then, is low-cost oil, or else it would not be produced! Even today oil remains in the ground in exhausted oil fields. This oil is a reserve for the future.

Moreover, one can assume that technology is not constant. Scientists and engineers will gain more knowledge of energy-producing systems and how to operate them. Throughout the world today, thousands of scientists and engineers are aggressively striving to develop new energy materials and producing systems. That is why oil companies prefer to be called energy companies. Future technology cannot be evaluated nor can the rate of change be known. If the past is any criterion, however, the future will be different from today.

Oil-producing and Oil-deficient States

The potential problems between the oil-producing and oil-deficient states primarily concern the regulation of output and of prices and the power to explore and develop the seabed. Many oil-producing nations such as Saudi Arabia are not industrialized, and they have had relatively low levels of per capita income. Oil riches have brought great changes in per capita income and the desire and ability to industrialize. At the same time, however, the real price of petroleum in the twenty-year period between 1945 and 1965 decreased as low-cost supplies increased faster than demand. National income was sustained by increases in output. Today cheap oil from the seabed would contribute to total world output and tend to lower prices. Petroleum produced from the seabed would then be a substitute for the production from existing oil-producing states and would affect both prices and output.

The oil-producing nations and others preparing to produce oil from the seabed are justifiably concerned about the future of this major source of income. Those fears have encouraged the growth of cartel-like organizations such as the Organization of Petroleum Exporting Countries. From time to time, calls are heard to control prices by developing a worldwide commodity agreement to regulate the output of petroleum. This plan is similar to that of the Texas Railroad Commission, which controls output in one part of the United States. Since cost is not controllable, cartels operate by manipulating output and prices. The objective is, of course, to maintain a "proper" and positive difference between revenue and cost. The result of this difference is a tax on the oil-consuming nations.

Over the long run, raw-material cartels cannot be maintained if the price is held above the price of the margin producer. New producers will enter the market and high-quality resources will become exhausted. In the short run, however, a cartel can be quite effective in controlling prices and output. Cartels also can be effective if they have broad governmental support and few members. Allocation of output becomes a severe problem as the number of members grows. If the geological assumptions about the oil in the seabed are correct, it is conceivable that many petroleum-producing and -exporting nations could exist. Since the cost of production would be different for most of these producers, it is unlikely that acceptable allocations of output could be made. In this event it is also doubtful that a successful cartel could be developed. The result could be price wars in petroleum sales.

Many of the potential seabed oil-producing states do not have the capability to finance and produce oil from their own wells. Although the productive-managerial personnel could be provided by a major private or state oil company, many producing nations desire not only to control their own resources, but also to develop the ability to manage and operate the wells. This desire is part of current nationalism. The major petroleum firms are in a weak position to resist this nationalistic force because major arguments could result in a complete takeover of production facilities rather than the partial nationalization that currently exists. Although years ago such a takeover was not possible, producing nations now have much greater power to enforce their will because of the energy crisis.

The oil-deficient nations are mainly industrialized countries that need assured supplies of energy to maintain their current levels of income and more energy if they wish to grow. Many of these nations face balance-of-payments problems as a result of deficits from importing oil. Presently most industrial petroleum comes from the Middle East states, which have growing negotiation power. However, the exporting region could become an instable supply source as a result of political factors. Additional and widely dispersed supplies would undoubtedly benefit the oil-deficient nations by increasing competitive marketing. The seabed offers opportunities for increasing supplies without major increases in cost. Therefore, the oil-deficient nations will probably take the international actions that would increase supplies by encouraging seabed exploitation; the oil-rich nations will probably take action to retard growth of seabed oil production.

Economic and Political Ramifications

Oil in the continental margins offers a new major supply source at current prices and potential for future production. In the short run, the price of oil may decrease somewhat as new output reaches the market. However, the growth of demand for energy seems to outpace the growing supply. Current known supplies of petroleum available to the United States at current prices are sufficient to meet anticipated demand. The National Petroleum Council states, "Existing reserves coupled with the non-communist world resource bases remaining to be discovered . . . are sufficient to meet requirements up to 1985."

Then what is the energy problem? There is a problem because political decisions are intermingled with economic decisions. A nation's policy for seabed exploitation depends on its vital interests. The United States policy was stated in a submission of the Draft Convention of August 1970 to the United Nations. This proposal called for prompt, efficient exploitation of the continental margins. The policy of oil-rich nations is to promote less prompt, efficiently controlled exploitation. Other nations have made other proposals. The essence of each proposal involves the world as each nation perceives it.

The United Nations established a working group on the international regime composed of thirty-three members. The result was the Declaration of Principles adopted by the General Assembly, Resolution 2749 (XXV), in 1970. Since then working groups attempting to translate principles into treaty articles have encountered many problems. Now it is planned that working groups will meet in Santiago, Chile, in the spring of 1974 to draft a treaty.

The treaty must be workable and provide rules for exploitation that do not increase the cost of production but do provide for safety practices, pollution controls, reporting procedures, and similar items. Rules that control output in order to control prices, however, should be seriously evaluated. Output controls must allow for a sufficient rate of production so that capital and interest is repaid, even at the lowest rate of production. This practice would reduce the financial risk of the state or private producing companies and give them access to capital sources.

Another area to be negotiated is the extent of ocean boundaries. There is no logical way to divide the seabed according to width, depth, and geological structure. A solution for developing seabed boundaries will undoubtedly be guided by some form of give-and-take.

A determination will have to be made of the amount of revenue that will accrue from seabed exploitation and how it should be divided among the nations of the world. If only a small amount of money is produced from seabed exploitation, there may not be a severe problem. The revenue could come from several sources: bidding on the right to explore and exploit, leasehold rents, production taxes, income taxes, and license fees. During the past twenty years of seabed leasing for oil exploration and production, the United States government has received $4.3 billion in lease bonuses, $13.1 million in royalties, $100 million in lease rentals, and some miscellaneous income for a total of $4.5 billion. These figures show the potential governmental income that could be developed from leasing and producing from the oil in the continental margin.

Many people think of mineral exploitation as a bonanza industry and expect great wealth to develop through petroleum exploitation. This view is not correct. Not all the seabed resources will provide rents available for appropriation. Many marginal oil fields will be found. Nonneutral taxes placed on marginal wells will force these wells to close.

Defining the duties of the Seabed Authority is the fourth major negotiating area. Will it only be a registry office or will it, as some members of the United Nations wish, be an organization capable of exploration and exploitation? There are, of course, many positions between those two extremes. The underlying questions are whether the organization will be deliberative, planning, operative, regulatory, or some combination of these functions.

Another item for negotiation is determining the powers and composition of the governing board of the Seabed Authority. The UN working groups will have to decide whether voting of the members will be weighted, who shall be included in the membership, if there will be permanent members of the board, and whether members of the board will include representatives from industry, engineering, and science. The board undoubtedly will determine the policy of the Authority and, over time, will be increasingly divorced from the membership of the United Nations itself. The Authority could become a semiautonomous group without control from the UN. The amount of control to be exercised by the UN is also a major question.

Enforcement and arbitration power of the Seabed Authority is the sixth area of negotiation. The international oil industry is especially concerned about this aspect of the treaty. The industry is fearful of expropriation without prompt, adequate, and effective compensation of production equipment and producing rights. On May 23, 1970, President Nixon enunciated five principles regarding exploitation of the seabed resources: (1) to collect substantial mineral royalties to be used for international community purposes; (2) to prevent unreasonable interference with other uses of the oceans; (3) to protect the ocean from pollution; (4) to assure the integrity of the investment necessary for such exploration; and (5) to provide for peaceful and compulsory settlement of disputes. In the president's policy the last two statements are clearly meant to refer to the oil industry's concerns.

The final negotiating point is preserving the interdependent ecology of the seabed. Disturbance of the bottom by oil exploration methods may decrease the quantity of fish in the waters above the continental margin. As on land, where surface mining disturbs farming and animal life that later affects human life, life in the sea is interdependent. The sea has many uses, and a disturbance caused by one use affects other uses.

These seven major negotiating points sum up the political, social, and economic concerns of the nations and will directly affect exploitation of the continental margins. The prime problem of the energy concern seems not to be a shortage of known reserves of petroleum, but, rather, the ancillary components of energy production. The continental margin waits to be exploited, but major negotiations are required among nations and oil-industry representatives. Until these international negotiations are completed, it may be anticipated that international petroleum firms will operate in waters adjacent to their own countries. In the United States, exploration of the continental margin will continue and will increasingly move toward deeper waters. Perhaps the push of technological change will increase the rate at which international negotiations will proceed.

International Implications

Oil and Middle East Politics

CHARLES ISSAWI

It will be assumed here that the huge increase in the world's demand for energy during the next ten or fifteen years can be met only by expanding the output of oil. Such an expansion would encounter great difficulties. In the advanced parts of the world, these difficulties are physical and have high financial and environmental costs. They include the gales and storms of the North Sea, the Arctic weather of Alaska and Athabasca, and the huge amounts of rock to be crushed and water to be pumped in order to extract oil from the shales of Colorado, Utah, and Wyoming. In the less developed regions, where physical conditions are more favorable and costs of extraction low, the difficulties are primarily political. These areas include Venezuela, Nigeria, North Africa, and above all the Middle East. During the 1970s and 1980s the bulk of the increment in oil production will have to come from the Middle East, more particularly Iran, Abu Dhabi, Iraq, and especially Saudi Arabia, which have huge developed and potential oil reserves.

The problems posed for United States foreign policy in the Middle East arise from the vortex of regional and local forces in the oil-producing countries, as well as from the impact of powerful forces originating outside the region. But before discussing these forces, it is necessary to review briefly the recent economic and financial development of the Middle East oil industry in order to indicate future trends and their implications for the United States and for the world.[1]

Developments in the Middle East Oil Industry

The development of the Middle East oil industry started before World War I. Until the early 1950s the region was politically and militarily dominated by Britain and, later, also by the United States. The world petroleum market was controlled by seven American and British companies that had full power in matters of production, pricing, and marketing. The Middle East governments merely received a fixed royalty averaging twenty to twenty-five cents a barrel.

The discontent and tension produced by World War II, the changes in Venezuela, the 1950 agreement in Saudi Arabia, and the Iranian nationalization crisis of 1951 set a new pattern. Instead of a fixed royalty, the governments received 50 percent of the profits made in producing operations, which included the extraction of oil and its delivery to tankers. In the 1950s, this averaged some 70 to 80 cents a barrel.

Since costs of production in the Middle East were low, the major factor determining profits became the "posted" price of oil. The Middle East governments became keenly interested in pricing. Hence the reductions in oil prices in 1959 and 1960, caused by increasing pressures in the market, disturbed the producing governments and led to the formation of the Organization of Petroleum Exporting Countries (OPEC). At present, OPEC includes Venezuela, Nigeria, Indonesia, and all the major oil-producing countries of the Middle East and North Africa. It accounts for over half the world's output of oil and for some 90 percent of oil exports. In its first ten years, OPEC failed to achieve its announced objective of restoring prices to their pre-1959 level, though it stopped further declines in posted prices. It did gain a series of minor though cumulatively important financial concessions, which together raised the average income of Middle East governments to 85 cents per barrel in 1969. More important, it accustomed these governments to working together and put them in an excellent position to

[1] For further details see Charles Issawi and M. Yeganeh, *The Economics of Middle Eastern Oil* (New York: Praeger, 1962); Charles Issawi, *Oil, the Middle East and the World*, The Washington Papers, vol. 4 (New York: Center for Strategic and International Studies, 1972); M. A. Adelman, *The World Petroleum Market* (New York: Johns Hopkins University Press, 1972); and James E. Akins, "The Oil Crisis," *Foreign Affairs* 51 (April 1973), 462-90.

take advantage of the unforeseen situation that arose in 1970.

The closure of the Suez Canal during the Arab-Israeli war of 1967 and the consequent strain on tankers had caused the companies to more than double their output in Libya, which was much closer to the European market. Thus Europe developed a great dependence on that country. In 1970 an unexpected shortage was created by the decline of supplies from Nigeria, because of the civil war, and from Saudi Arabia, because of the damaging of the pipeline carrying its oil through Syria. The revolutionary government of Libya took advantage of the situation to put pressure on the companies operating in that country, and succeeded in raising both the rate of income tax levied on the companies and the posted price of oil. These raises in turn made it possible for OPEC, in its Tehran meeting of January 1971, to score its greatest success. The Persian Gulf countries obtained an increase in their share of profits to 55 percent and in posted prices of 35 to 40 cents, to be raised about 5 cents a year until 1975. In return the governments would not make any new demands until 1975, but they did insist on being compensated for the devaluation of the dollar. By now their average receipt per barrel is about $1.50.

Two further developments in 1971-73 should be noted. The entire oil industry in Algeria, most of it in Iraq, and some companies in Libya have been nationalized. Arrangements have also been made for "participation" with Saudi Arabia, Qatar, and Abu Dhabi, under which these countries will acquire an increasing share of the capital of the companies operating there, achieving majority control by 1982. Various declarations by the government of Iran have made it clear that the major concession in that country will be considered to have lapsed by 1979 or at the latest 1984.

Developments of the past five years indicate trends that may be anticipated during the next five or ten. Nationalization, which until recently was regarded as the main danger, is of course still possible. Much will depend on how much profit Iraq will get from its nationalized industry compared to, say, Iran or Saudi Arabia. But the most probable and worrisome trends are the rapidly escalating prices of oil and a declining willingness of the main producing countries to supply the huge amounts that will be demanded in this decade and the next. Of course, these two forces can reinforce each other.

Compared to its extremely low cost of production, some 10 to 20 cents a barrel, Middle East oil is priced high, at present $2.70. By any other measure, however, it is still cheap. Between 1948 and 1973 its price rose by some 35 percent, an average of less than 1.5 percent a year, which is low compared to most commodities. Compared to the price charged for petroleum products in consuming countries, the cost of Middle East *crude* oil is very low. Thus in 1970 the average sales taxes levied on petroleum products of six major countries in Western Europe was about $7.53 per barrel-equivalent or some five times what is being received by the producing governments of the Middle East.[2] Lastly, compared to other sources of energy, such as coal, gas, and nuclear power—and, of course, oil produced in any other part of the world—Middle East oil is still very cheap. There is, then, every reason to believe that if world demand for oil continues to grow fairly fast, the OPEC countries will raise their prices steadily and perhaps even sharply. Such a raise will affect the balance of payments of the importing countries, of which the United States has become one of the most important. Recent calculations suggest that by 1980 oil imports may be costing the United States over $20 billion a year and Western Europe and Japan over $40 billion, compared to $2 billion and $11 billion in 1970.

But there is another more disturbing possibility. The major producing countries may deliberately curtail their output of oil. Indeed, some have already done so. Libya has drastically cut back its production while Kuwait and Venezuela have stabilized theirs. There are numerous interconnected reasons for such policies. The recent monetary instability, particularly the weakness of the dollar, has made the holding of cash reserves much less attractive than in the past. This instability is made worse by the powerful inflationary pressures operating throughout the world, which reduce the value of present earnings and make it increasingly desirable to spread out production and postpone as long as possible the depletion of oil reserves. It is true that in the coming decades oil will face increasing competition from other sources of energy, and it is probable that well before the end of this century it will have lost its leading place to nuclear power or some other new source. Even so, a market for all the world's oil is assured for the foreseeable future, as a feedstock for petrochemicals if not as a fuel. Because of the low-price elasticity of demand for oil, a restriction in supply should lead to higher prices and increased total receipts.

[2] M. S. Al-Mahdi, "The Pricing of Crude Oil in the International Market," Eighth Arab Petroleum Congress, Algiers, May 28-June 3, 1972.

Another important factor is that most of the major oil-producing countries already have huge reserves of foreign exchange—amounting to billions of dollars in the cases of Libya, Saudi Arabia, Kuwait, and Venezuela. These reserves are bound to be swollen by the great increase in oil revenues, which for the Middle East and North African countries alone may reach $40 billion (1973 value) annually by 1975. Although some of the main producers, such as Iran, Venezuela, and Iraq, could profitably invest all their potential oil earnings at home, the smaller countries could not. It may not be in the economic interest of the smaller countries to increase production at the rate demanded by the rest of the world, which will have to provide the required motivation for expansion.[3] This motivation could be provided by enlarging these countries' absorptive capacity through a massive infusion of technical and managerial know-how and also by inducing them to invest heavily outside their borders. Such investments could be made in neighboring and other underdeveloped countries that are short of capital, in the petroleum industry itself, and in other branches of the economy in advanced countries.

Political Issues Within the Region

The problems for United States policy in the Middle East are caused by disruptive forces originating within the region and forces impinging on it from outside. The disruptive forces within the region may be divided into four categories: internal revolutions, territorial disputes, national rivalries, and ideological antagonisms.

Revolution is endemic in practically the whole Third World, and the oil-producing countries are no exception. Of course, oil acts as a lubricant, making it possible to raise living standards rapidly and to extend social services to large sections of the population. But, as is being increasingly realized, economic growth is a deeply unsettling process. It almost inevitably erodes the social and moral bases on which a community has stood since time immemorial. More particularly, economic development almost inevitably leads to greater concentration of wealth and a widening of the gap between rich and poor. This disparity often results in a great increase in corruption at the top. Such a process is

[3] Even countries that need capital very badly may decide to wait: "Algeria will become a world leader in the export of gas. But unless the right financial terms for deals can be agreed, the Algerians will just sit on their gas until the would-be purchasers change their tune." *The Financial Times* (London), June 27, 1973.

both accelerated and magnified when wealth comes quickly, as has happened in the oil-producing countries. The very real achievements of the monarchies in Iraq and Libya were not sufficient to offset the resulting st ains, and the revolutions in these countries will surely not be the last in the region.

A revolution, though disturbing, does not have to be disastrous for American interests. The new government is at least as dependent on oil revenues as the old and, after the initial fervor has cooled off, tends to take a pragmatic view of its own national interests. In this respect, the example of Algeria is both significant and encouraging. But since the United States and the oil companies have to deal with and support existing governments, they become identified with the old regime and are inevitably regarded with suspicion and hostility by the succeeding revolutionary government. In these circumstances, the United States has to carry out the diplomatic equivalent of a backward somersault, and the performance, though not impossible, calls for great agility. Such agility is going to be increasingly needed, since the probability of the overthrow of the existing regimes in all the oil countries is high, although in one or two—notably Iran—the political and social structure may by now be strong enough to stand.

Territorial disputes within the region are numerous. In the area of the Persian Gulf, many of them are very old, but all were frozen during the period of British domination of the gulf and emerged only when it became clear that Britain was about to withdraw and leave the local states to maneuver for position. Iran, for example, claims the three tiny islands at the mouth of the gulf, Abu Musa, and the two Tumb islands. Iran occupied these islands in November 1971, with almost no casualties, but the incident was used by a country as distant as Libya to expropriate a British oil company!

A much more serious dispute is that between Iran and Iraq over the Shatt al-Arab River, which forms part of the boundary between the two states, constitutes Iraq's outlet to the sea, and provides access to Iran's two main ports, Khorramshahr and Abadan. Iraq has also laid claim to Kuwait; its attempt to enforce its claim in 1961 was thwarted by the other Arab states, led by Egypt and Saudi Arabia, but it is now being revived. The ill-defined boundary between Saudi Arabia and Abu Dhabi has given rise to many controversies and some armed clashes; the stakes consist of billions of barrels of probable oil reserves. The frontiers between other Arab sheikhdoms are also ill-defined, and those between Saudi Arabia, Oman, Yemen, and South Yemen leave large areas of contention.

Outside the gulf, the main territorial disputes are ones between Israel and its Arab neighbors, which, however, are properly regarded as a national struggle. There are also latent disputes between Syria and Turkey over Alexandretta and the border areas. National rivalries, sometimes erupting into struggles, exist between the five main ethnic groups inhabiting the Middle East: Arabs, Turks, Iranians, Kurds, and Israelis. The Kurds have at various times risen against the Turkish, Iranian, and Iraqi governments, among whom their country is partitioned; at present their main effort is in Iraq, where fighting may flare up again. Generally speaking, a common interest in preventing a Kurdish uprising has drawn the three governments together, but at various times one or another of them has supported an uprising in a neighboring state. Thus, at the moment, the Kurds in Iraq seem to be getting considerable help from Iran.

Turks and Arabs have not engaged in any armed conflict since the latter gained their independence following World War I. But border disputes have stirred animosities. The cold war and differences in approaches to Israel have often led to tensions. At present, however, relations between the two peoples are probably better than they have been for a long time.

The Arab-Iranian rivalry for hegemony in the gulf developed recently, and it is increasing in intensity. At various times Iran has been confronted by Egypt and Saudi Arabia, but now its main antagonist is Iraq. The latter, surrounded by unfriendly neighbors, has sought to strengthen its position by drawing closer to India in opposition to Iran and its close ally Pakistan. The Soviet Union has supported Iraq, while the United States has helped Iran and Pakistan.

The Arab-Israeli conflict is the most serious and intractable in the region and the one with the widest ramifications. It opened a way for Soviet penetration into Egypt and Syria, thus affecting the strategic and political balance in the Mediterranean. By dislocating a million Palestinian Arab refugees who see no hope for themselves within the existing order, it has generated the most powerful revolutionary force undermining the Arab governments, particularly those friendly to the United States, such as Jordan and Lebanon.

The effects of the Arab-Israeli conflict on the supply of oil, though indirect, are great and deeply disturbing. During both the 1956 and

1967 wars the flow of oil to the West was temporarily halted, partly because of deliberate policies of Arab governments and partly because of spontaneous disruption. Both types of action may be expected to recur, probably on a larger scale, in the event of another conflict, but this time the West is in a far weaker position to meet the crisis. For one thing, dependence on Middle East and North African oil has greatly increased, and the large reserve capacity that the United States and Venezuela had, and which was used to maintain supplies, no longer exists. There are fewer friendly governments that can be called upon to increase their output to offset declines elsewhere, as Libya did in 1967.

The Arab governments, including the one that is most important in this contest, Saudi Arabia, have repeated with increasing emphasis their intention to use oil as a weapon to bring about a change in American and European policy toward the Arab-Israeli conflict. This threat should not be lightly dismissed. No less serious is another potential weapon in Arab hands—the vast monetary reserves that will be accumulated over the next few years by the oil-producing countries and that could threaten the stability of one or more Western currencies.

The situation presents the United States government with grave dilemmas. Should it seek to conciliate the Arabs, who though increasingly resentful are not yet completely alienated and would probably respond to some positive moves on its part? Should it despair of Arab cooperation and base its policies on total commitment to Israel and Iran? Should it go further and make it clear that a stoppage of oil would be met by military countermeasures? All these policies have their advocates and all can be more or less persuasively defended, but it seems unlikely that any of them will be wholeheartedly pursued to the exclusion of the others. The outside observer can only urge that another powerful effort be made to settle at least some of the outstanding Arab-Israeli issues, preferably by concentrating on the one that seems most amenable to a solution, that between Egypt and Israel.

Finally, there are the ideological antagonisms, usually considered a conflict between "conservative" and "progressive" forces. This distinction is rather meaningless if applied to the domestic policies of the two groups, but it does denote opposite attitudes toward revolution. The conservative states are the large oil producers—Iran, Saudi Arabia, and Kuwait; the small oil sheikhdoms; Jordan, Lebanon, and Yemen; and, in North Africa, Tunisia and Morocco. On the other side—but by no means working in harmony—are Egypt, Iraq, Syria, Algeria, and Libya, which in various ways are putting pressure on the conservative states. South Yemen is in a category apart. Its main activity is to sustain the rebellion against the sultan of Oman, with the hope of eventually spreading revolution to the whole Arabian side of the gulf. The conflict between conservatives and revolutionaries—the former long on wealth, the latter on ideology—is bound to be intense and prolonged, and adds one more element to the Middle East turmoil.

The United States and the Other Powers

United States policy is entangled with those of all the other major powers, mainly though by no means exclusively because of oil. First, there are its European allies, particularly Britain, France, West Germany, and Italy. In judging current United States-European relations, it should not be forgotten that the United States is a newcomer in the Middle East. Until after World War II the region was to all intents and purposes a British preserve, with France playing a secondary part. Similarly, the Middle East oil industry was overwhelmingly controlled by British capital, with American and French companies holding minority shares. In the 1950s, American capital became predominant in the oil industry, and the United States became the major guardian of Western strategic and political interests. The Europeans accepted this change, though not without resentment, particularly when the United States used its power to coerce its allies, as it did during the Suez crisis of 1956. But as long as Europe was weak and disunited and the United States ensured the defense of the Mediterranean region and the supply of oil, Europeans had little choice but to acquiesce and cooperate.

Now, however, Europeans are becoming seriously concerned about their future supplies of oil from the Middle East and North Africa, which currently account for 60 percent and 20 percent of West European imports. They are alarmed at the prospect of separate deals between the United States and Saudi Arabia, and perhaps between the United States and Iran, which would give the United States a privileged position in the marketing of oil from those countries and, in a time of scarcity, ensure that its needs are met, if necessary, at the expense of Europe. There has been much talk of such deals and indications that they might be welcome to some Middle East exporters. Europeans are worried by the possible effect of present United States policies in the

Middle East on Western oil supplies. More specifically, they fear that the pro-Israel policy of the United States will result in either an Arab embargo or a disruption that could stop the flow of oil.

If the Europeans could act as a bloc, they would be in a very strong position in this as in many other matters. But at present each major country believes it is in an advantageous position and therefore prefers to play a lone hand—Britain because of its international oil companies and North Sea deposits, France because of its good relations with the Arabs, and Germany because of its great economic strength and lack of a colonialist past in the Middle East, and so on. Hence, while Britain seems to be following the American lead, France has negotiated separate agreements with Algeria and Iraq, Italy is continuing the policy initiated by the Ente Nazionale Idrocarburi in the 1950s of breaking into various areas, and Germany and other European countries have set up national companies which are seeking concessions in the Middle East and North Africa. Competition between European and American companies may have far-reaching consequences in the oil market and may also strain relations between governments. And it must not be forgotten that oil, though important, is only one strand in the complex web of economic, political, and strategic cooperation and competition binding Europe and the United States.

This analysis also applies to Japan, which obtains nearly 80 percent of its oil from the Middle East. It too has depended entirely on the United States to safeguard its supplies, and it too is becoming alarmed. Japan is in no position to exert any political pressure in the Middle East, but it has much to offer economically. Hence it too is seeking to obtain new concessions in Abu Dhabi, for example, and to dissociate itself to a certain degree from the United States.[4] It is also making every effort to diversify its sources of supply by developing oil in Indonesia, Siberia, and offshore, while realizing quite clearly that for a long time to come its dependence on the Middle East will continue to be considerable.

The policy of the People's Republic of China in the Middle East, on the other hand, seems not to be in any way influenced by oil, in which it is at present self-sufficient. China's growing involvement in the region, including an expanding volume of trade with most countries and a small amount of economic aid to some, seems to have two aims. First,

[4] *The Financial Times* (London), July 16, 1973.

it is trying to help such revolutionary movements as the guerrillas in Oman and the Palestinian Arabs by providing weapons and training. Second, and probably more important, it seeks to prevent the extension of Soviet influence in the Middle East and Indian Ocean areas. Hence the paradoxical spectacle of its excellent relations with regimes commonly regarded as reactionary, e.g., Iran and Pakistan.

India has, since its independence, enjoyed the friendship and goodwill of the Middle East, but the recent war in Bangladesh put a strain on its relations with all Muslims, including those in the Middle East. At present, its main diplomatic thrust is to get closer to Iraq, partly to ensure future oil supplies. To some extent it has a strategic and political aim: cooperation against Iran and Pakistan, which are aligned with the United States and China respectively. This move surely has been encouraged, if not inspired, by India's new ally, the Soviet Union.

Soviet influence in the Middle East has greatly expanded in the past fifteen years and may be expected to grow further. It now takes the form of access to naval and air bases, supply of arms, treaties of alliance, predominant influence in several important countries, notably Egypt, Syria, and Iraq, and a foothold in the oil industry. The cost of all this to the Soviets has been high, running into very many billions of dollars—and rapidly increasing unpopularity in the region—but not necessarily excessive given the magnitude of the stakes.

There have, of course, also been some important reverses. The Middle East policy of the Soviet Union is part of a large and complex whole. One main element seems to be fear of China, with a consequent desire to establish a commanding position in the Indian Ocean. Another is a détente with the United States that would lead to a withdrawal of American troops from—and a weakening of American influence in —Europe, thus opening the way for Soviet predominance in that continent—and at the same time greater access to American and European capital and technology.

The weakening of United States influence in the Middle East is part of this design, which has to a large extent been already achieved by backing local nationalist and radical forces in their struggle against Western economic and political positions in the area and supporting the Arabs against Israel. But the Soviets also seem anxious to avoid a direct clash with the United States or even one by proxy through another Arab-Israeli war. As has been aptly said, they want the Middle East pot to simmer, not boil over. Moreover, their very involvement has limited their freedom of action. For example, they encourage Iraq in its activities in the gulf, but not to the point of ruining their relations with Iran.

As for oil, the Soviet Union's aim is first and foremost to end Western control. Any rise in oil prices that may occur would be to their advantage by handicapping Western industry. At the same time, they are slowly increasing their imports of Middle East oil, partly to meet the rapidly growing needs of the East Europeans and partly to free some of their own oil for export to Western markets in exchange for highly desirable machinery and technology. Such a policy runs counter to the interests of the Middle East oil producers, who increasingly refuse to deliver oil in exchange for Soviet goods and demand convertible foreign currency. If the nationalist sentiments of the oil producers point in the direction of the USSR, their economic interests point to the West, and their attempts to secure both political and economic objectives are often incompatible.

Thus the United States is confronted with a complex situation in the Middle East, full of tensions and contradictions. The implementation of United States policy over the next few years will demand statesmanship and skill of the highest order, and the government will need much support from the American public in its difficult task.

Energy Self-sufficiency and National Security

JOAN EDELMAN SPERO

Self-sufficiency in energy supply has been a tenet of American national-security policy for the past twenty years. This policy of self-sufficiency has been based on a broad definition of national security, which includes not only military defense but also the economic strength that underlies military defense. Thus national security means protecting the United States against military threats such as nuclear war, limited war, and general, nonnuclear war and against economic threats that would significantly disrupt the economy.

The problem of maintaining sufficient energy varies with the type of security threat. In the event of a massive nuclear confrontation, domestic production of energy would be sufficient to meet the needs of a shattered economy.

In a limited war, import interruptions of energy would also be a minor concern. Potential supply interruptions in such a war would be of two types: embargoes by producing states and forced supply interruption by an adversary. An effective embargo is not likely to occur in a limited war because such a war probably would not affect the vital interests of producing states. Limited wars that do not directly involve producing states—as in Korea and Vietnam—do not pose a threat. Only a limited war in which producing states were participants or closely allied with participants would result in an embargo.

Forced interruption by an adversary is also unlikely. The only potential adversary capable of such interruption is the Soviet Union, but any conflict with the Soviet Union would not be limited. In any case, the Department of Defense maintains plans to supply petroleum for military requirements from United States sources if foreign sources are interrupted by a limited war. Civilian supplies would most likely not be a problem.

The real national-security problem would occur in the event of a general, nonnuclear war of long duration. Enemy military activity would almost certainly be directed at supply interruption either by destruction of production facilities or blocking of supply lines. In addition, alternate supply through reexport from friendly countries would be unlikely because enemy activity would be aimed at eliminating all sources of supply.

Self-sufficiency can be crucial in a long, general, nonnuclear war. In World War II, secure American oil supplies proved to be a key military advantage while Axis fuel-supply problems were a determining factor in their defeat. Today, like the situation during World War II, the oil supply is sufficient to ensure the functioning of American armed forces in the event of a general, conventional war. Military requirements today are such a small fraction of total domestic consumption that oil for the armed forces is unlikely to be jeopardized. At peak levels of World War II, military consumption took 33 percent of total national demand. The Department of Defense predicts a maximum total military demand of 10 percent of current consumption. By 1980, that figure will be 7 percent.[1] The real problem today would be a threat to the functioning of the national economy, of defense production, and of essential civilian uses such as utilities, space heating, and motor vehicles. A general, nonnuclear war is considered a possibility but not a probability or the key risk to national security. Energy-security planners consider it unlikely that such a war would continue for more than a few months without settlement or escalation. However, in assessing national-security energy problems and in planning energy policy, such a contingency must be considered.

The more probable security threats are nonmilitary. An economic threat to national security does not involve problems such as rising energy costs, injury to particular domestic industries, or even damage to the United States's competitive position in international trade. But a vital threat to the functioning of the economic system as a whole

---

[1] U.S., Congress, Senate, Committee on Interior and Insular Affairs, *Oil and Gas Imports Issues*: Hearings on S. 45, 93d Cong., 1st sess., January 10, 11, and 13, 1973, p. 745.

would be a threat to national security. A major interruption of the energy supply that crippled the economy would pose such a threat.

Energy dependence may also have political consequences. Military and economic vulnerability decreases the United States's international bargaining position and makes it vulnerable to political pressure and political blackmail. Threatened or actual supply interruptions, for example, are effective bargaining tools on the side of producer states and their allies. Such political pressure threatening vital American interests would be a national-security problem.

Policy of Self-sufficiency

National-security planners view self-sufficiency as a central means of avoiding threats posed by dependence on foreign sources for energy. Oil has been the only major energy source that threatened the policy of self-sufficiency. Until 1949, the United States was a net exporter of oil; however, with the growth of domestic energy consumption and the development of inexpensive foreign oil, the United States in the 1950s developed into a net importer of oil, causing the government to become concerned about oil dependency. The statutory authority for oil-import controls is the Trade Agreements Extension Acts of 1955 and 1958 and the Trade Expansion Act of 1962, which permit the president to restrict imports that threaten to impair national security. In 1955 a cabinet advisory committee recommended for the first time the use of voluntary restraints to maintain the 1954 ratio of crude and residual-fuel imports to domestic production. In 1957 the Eisenhower administration recommended continued and voluntary restraint on imports. Finally, in March 1959, a presidential proclamation established the Mandatory Oil Import Program. This program with various minor revisions operated until April 1973.

The purpose of the oil-import program has been to encourage domestic oil production by protecting domestic producers from low-cost foreign imports. Quotas were set on oil and oil products coming from foreign sources. Canada and Mexico were excluded because their oil exports did not depend on ocean transport and thus involved no national-security risk. Venezuela, whose oil was shipped by tanker, was given preferential treatment because supply lines in the Western Hemisphere were relatively secure. Based on total American consumption of oil, quotas were set at roughly the level of imports in 1959, about one-eighth of total consumption.

The rationale behind the oil-import program and the policy of self-sufficiency in general is that the only way to assure domestic supplies of oil adequate for national security is to use American oil fields. To ensure that the drilling and exploration necessary for self-sufficiency is carried out, domestic production must be protected from foreign competition. Thus, as one observer put it, the policy has been to safeguard the American supply by using it up.

This rationale has strong support from the American petroleum industry, from the independents (companies whose principal activities are in the United States) as well as from the internationals (companies with both domestic and international activities). The independents' very existence has been preserved by the oil-import program. The internationals have benefited from their domestic production because of the program and have found adequate markets outside the United States for their international production.[2]

Other national energy policies have supplemented the oil-import program: special tax treatment for oil and gas that encourages domestic production; research and development programs in coal and nuclear energy that contribute to the development of alternate domestic forms of energy. However, none of these policies have been specifically for national security.

As recently as February 1970, President Nixon's Cabinet Task Force on Oil Imports predicted that present measures, or similar programs that might replace them, would ensure American self-sufficiency in the foreseeable future. Warning of the danger to national security of importing large quantities of oil from outside the Western Hemisphere, the task force set 10 percent as the peril point for energy imports from the Eastern Hemisphere.

Only three years later, however, the policy of self-sufficiency collapsed. Since April 1973, when President Nixon announced the removal of import controls on oil, import levels from the Eastern Hemisphere have soared well beyond 10 percent. Self-sufficiency was a victim of rapidly increasing domestic energy production. Most experts predict that the current trends, which created the energy crisis, will continue until 1985.

[2] Berhard Brodie, *War and Politics* (New York: Macmillan Co., 1973), pp. 296-98; idem, *Foreign Oil and American Security*, Yale Institute for International Studies Memorandum no. 23, September 1947; and Robert Engler, *The Politics of Oil* (Chicago: University of Chicago Press, 1961), chap. 9.

Thus, as the energy gap increases, the United States will become gradually more dependent on foreign sources of oil and, to a lesser extent, on natural gas. The United States will remain self-sufficient in coal, hydrogeothermal energy, and nuclear energy. But an increasing percentage of oil, which will make up nearly half of total energy consumption in 1985, and natural gas, which will make up about one-fifth of consumption at that time, will be imported. By 1985 the United States will be importing about 50 percent of its oil needs and 15 to 16 percent of its natural-gas needs. Thus, as in the past, oil will be the major energy import and the major national-security problem. What national-security problems does this end of self-sufficiency pose? How can the United States deal with the changed energy-security situation now and in the future?

Geography and Politics of Oil Dependency

The principal source of American oil imports up to the present has been the Western Hemisphere, particularly Venezuela and Canada. Since Western Hemisphere sources have friendly governments and can ship supplies with little risk of interference, the United States views these sources as secure. Apparently, however, the ability of these countries to supply the growing United States import needs will be limited in the near future.

Production in Venezuela seems to have stabilized, and Venezuelan oil has decreased as a percentage of total United States imports. Venezuelan reserves are down; and, in spite of prospecting, no new sources have been found. Moreover, Venezuelan oil with its high sulfur content is less desirable from an environmental viewpoint.

The prospect for Canadian oil is somewhat brighter. The percentage of imports from Canada has increased in recent years and should continue to rise. Significant reserves have been found in the Mackenzie Delta, the Arctic Islands, and off the east coast. There are also huge quantities of oil in the Athabasca tar sands. Canadian imports, however, will not increase sufficiently to fill a major portion of the energy gap between now and 1985. Although American and Canadian companies are eager to develop these resources, the lead time for exploration and development is long. Moreover, rising Canadian nationalism may lead to restrictions on increased exports of oil and gas to the United States.

There are several new sources of oil in the Eastern Hemisphere. Nigeria and Indonesia are now significant oil producers. Imports from these two regions would be militarily vulnerable because of long, sea-borne supply lines and may be politically vulnerable because of internal instability. In any case, Nigerian reserves are small, and Indonesian reserves, though larger, will be exported primarily to Japan and the Far East. Other potential sources in sub-Saharan Africa and the Far East will not be developed in the near future.

Another Eastern Hemisphere energy source is the Soviet Union. The USSR is at present an exporter of oil, though this situation may change as Soviet energy needs expand. However, any significant dependence on Soviet energy would pose a grave national-security risk. As the principal American adversary, the Soviet Union would be able to deny the United States energy in wartime and to use embargo as an effective bargaining tool. While the United States may make some energy arrangements with the Soviet Union, such as the recent natural-gas agreements, the Soviet Union cannot be considered a dependable energy source.

Since the Middle East and North Africa possess the world's largest oil reserves, the greatest prospects for future discoveries, the cleanest oil, and the lowest production costs, this region will be the major source of American oil imports in the future. Government and industry experts now predict that by 1980 one-third or more of total American oil consumption will come from the Middle East and North Africa, mainly from the Persian Gulf states of Iran, Saudi Arabia, and Kuwait.

Potential Security Problems

A high level of dependence on Middle East energy sources poses several potential security problems for the United States. In a general, nonnuclear war of long duration, dependence on oil supplies from this region could be a severe military handicap. The long, sea-borne supply line—especially long now because of the closing of the Suez Canal—would be vulnerable to enemy interruption. Much of the route passes through areas where American forces have little operating experience and few bases. In addition, tankers from the Persian Gulf must pass through the Strait of Hormuz, which can easily be mined or blocked. According to Admiral Elmo R. Zumwalt, Jr., chief of naval operations, sinking a handful of supertankers in critical passages could effectively block shipments from the gulf for a long time. There is little the United States could do to prevent such actions. In the event of an extended, general, nonnuclear war, it must be expected that the Soviet Union would attack American oil transports. The Soviet Union has a growing capability to interdict supply lines. It possesses a large submarine fleet which is increasingly nuclear-powered, a surface fleet about the same size as that of the United States, a developing sea-based air capability, and many antiship cruise missiles on submarines, surface ships, and aircraft. Defense of American and allied supply lines would thus require a significant military effort. Even in the case of successful defense of sea routes, supplies would be decreased as a result of reduced shipping capacity and the need for convoys.[3]

Finally, dependence on Middle East oil might increase the risk of escalation. American oil sources in the Middle East would be a tempting target for Soviet attack. But, if the Soviet Union succumbed to the temptation to strike oil facilities, it is uncertain how long the war could continue without American counteraction—perhaps the destruction of Soviet oil facilities—and the greater likelihood of escalation into nuclear war. To be sure, such a long-term, nonnuclear war is considered a relatively remote possibility by security planners. Yet, because of the seriousness of the consequences, this contingency must be considered.

Dependence on Middle East oil also poses threats to the vital functioning of the American economy and to the American international political position in peacetime. The most serious threat would be a concerted, politically motivated oil embargo. An embargo might be imposed to "punish" the United States for policies such as the support of Israel; or it might be used to pressure the United States to adopt a foreign policy more favorable to the producing states. A general embargo for political reasons occurred for one month, in 1967 at the time of the Arab-Israeli war when Arab countries accused the United States of participating with the Israeli air force in attacking Cairo. The embargo was lifted after the charges were proven false.

The 1967 embargo posed no serious security threat to the United States. Other non-Arab sources, such as Iran, were able to meet American needs. But with a greater level of dependence, an embargo is a more

[3] *Oil and Gas Imports Issues*, pp. 748-76.

serious threat. Any large-scale, long-term interruption of oil imports from the Middle East would significantly disrupt the American economy and would provide a powerful bargaining tool for states imposing the embargo.

The Arab-Israeli conflict of 1973 led to a series of politically motivated embargoes and reduced production aimed at changing the United States's pro-Israel policy and terminating American military assistance to Israel. The Arab oil-producing states announced total embargoes on oil shipments to the United States and reduced total oil production pending a return to normalcy in the area and an Israeli withdrawal to pre-1967 boundaries. These reductions were aimed at the United States's oil-consuming allies in Western Europe and Japan, making it more difficult for them to reexport Arab oil to the United States.

The willingness and ability of oil-producing states to institute a general embargo has obviously been increased. The Arab-Israeli conflict has acted as an important unifying force. Even more conservative and geographically remote states, such as Saudi Arabia and Kuwait, are increasingly responsive to anti-Israeli sentiment and to pressure from other Arab states to take action against the United States. Also important is the background of increasing economic cooperation of Middle East and North African oil-producing states through the Organization of Petroleum Exporting Countries (OPEC). This cooperation has led to successful bargaining for new economic agreements between the producers and foreign-owned oil companies and demonstrated the potential power of cooperation in the political arena. Finally, oil-producing countries have vast financial reserves that enable them to bear the economic cost of an embargo.

Because of the relatively limited degree of United States dependence on Middle East oil—about 10 percent of domestic consumption—the 1973 embargo has not endangered American national security. Depending on its length, the embargo will result in varying degrees of civilian inconvenience and discomfort and some economic disruption but no significant disruption of the economy is foreseen. The embargo of 1973, however, highlighted the problem of future United States vulnerability to the Arab oil weapon.

While a general political embargo seems increasingly possible, there are several factors that limit its likelihood and effectiveness. Not all Middle East oil-producing countries are Arab and involved in the Arab-Israeli conflict. Iran, one of the major producers, continued to ship oil to the United States during the 1967 and the 1973 embargoes. Within the Arab world, policies toward the United States differ. The present governments of Libya and Iraq would probably support a long-term embargo. But the present governments of Saudi Arabia and Kuwait, the major American suppliers, are conservative and have followed a more moderate policy toward the United States.

To be effective, an embargo would have to include all major importing countries. Unless Western Europe and Japan were included, oil could be shipped there and reexported to the United States. Thus, an effective embargo would have to include both areas and would be much more difficult to organize and maintain.

Finally, there is some question about the ability and willingness of producing countries to bear the economic cost of an embargo for a long period of time. The 1973 embargo and reduced production coincided with costly foreign-exchange outlays for military purchases from the Soviet Union. Arab states were forced to secure foreign loans to finance the purchase of military equipment for Egyptian and Syrian armies. How long producing states can afford to maintain embargoes and reduced production under these conditions is unclear.

Aside from a general and concerted political embargo, there is a threat of limited interruption of supply. With a tight market and limited alternate sources of supply outside the Middle East and North Africa, the danger of an oil shortage caused by the loss of the oil of one or two major producers in that region is significant. One State Department official predicts that the loss of production of any one of the major producers could cause a temporary but significant crisis. This means that the loss of two or more major sources could pose a vital economic national-security threat.

Several types of limited interruptions might occur. There is the possibility of physical interruption of supply because of local or regional hostilities, revolution, or guerrilla activities. Although guerrilla activity or revolution is not likely to interrupt the bulk of exports and regional wars are not likely to be prolonged, the possibility of such interruptions must be taken into account in planning.

In addition to politically motivated interruptions, there is the possibility of embargoes to raise the price of oil. Although such an embargo would cause no direct military or foreign-policy problems, it might threaten the national security if it severely strained the economy.

There is also the problem of nationalization of Western-owned oil companies operating in the Middle East and North Africa. In one form or another, such takeovers seem inevitable. Radical regimes can be expected to effect such outright nationalizations as Libya did with British Petroleum and Bunker Hunt. Even the most conservative regimes like Iran have indicated a commitment to increasing national control. However, such nationalizations do not pose a security threat to the United States. To be sure, nationalized companies will be less receptive than American-owned companies to American economic and political concerns. Nevertheless, nationalized companies will continue to find their major markets in the United States, Western Europe, and Japan, and the leaders of these producing countries recognize that fact. Thus, nationalization need not be a threat.

Finally, there are security problems posed by the role of the Soviet Union in the Middle East and North Africa. The Soviet Union recognizes that control of Middle East oil would be an important power advantage in its dealings with the West. This interest in increasing Soviet power *vis-à-vis* Western countries that depend on Middle East oil has been one of the principal motives for increased Soviet political, economic, and military activity in the Middle East. Growing Soviet influence in that area has raised the spectre of Soviet control of American oil supplies. The Soviet Union has developed its influence in the Middle East and North Africa by providing economic and military assistance to many Arab states, by supporting the Arab position in the Arab-Israeli conflict, and by backing various acts against Western oil companies. The USSR has increased its naval presence in the Mediterranean, the Indian Ocean, and the Persian Gulf. And the Soviet Union has made agreements with Iran, Iraq, Syria, Egypt, and Libya for the development and trade of oil and gas resources.

What does this mean for American national security? There seems to be very little likelihood that the Soviet Union will be able to shut off American oil supplies short of military action. No Middle East country, no matter how anti-American, would be willing peacefully to permit Soviet control of this vital national resource. The USSR could not sustain the economic cost of buying up energy that might otherwise go to the United States. Thus, the only way the Soviet Union could obtain control of Middle East and North African oil resources would be through direct invasion or interference with sea transportation of oil. But such force would meet American resistance. The more likely possibility of Soviet influence on American oil supplies is an increasing number of Soviet agreements with national oil companies. As Soviet energy-demands increase, the Soviet Union might emerge as a competitor for Middle East and North African oil. Nevertheless, for the foreseeable future, major oil markets for producing states will continue to be in the West.

Thus, the end of American self-sufficiency does not create a national-security crisis. In wartime, the only security problem posed by oil dependence would occur in a much prolonged, general, nonnuclear war. Although the consequences of dependency in such an event would be severe, the contingency seems unlikely. In peacetime, the major threats—general and concerted political embargo or Soviet control of oil supply and transportation—seem unlikely. Less significant but important threats seem more likely: physical disruption and limited political and economic embargo.[4]

Future Energy-security Policy

Must the United States completely abandon the traditional policy of self-sufficiency as impossible and unrealistic and accept the changed national-security position as immutable? Even the most vociferous advocates of self-sufficiency agree that such a policy would be untenable in the period up until 1985. All the United States can do is to minimize the risks of dependency. There are numerous steps the United States can take in this direction. It can attempt to decrease domestic consumption by using energy more efficiently and by adopting conservation measures. Smaller, more efficient automobiles and mass transit are examples of such measures. These policies, while helpful, cannot be expected to produce a reversal in consumption patterns. The Office of Emergency Preparedness, however, estimates that in a crisis a severe program of mandatory reductions could reduce petroleum consumption by as much as 30 percent.[5]

The United States can also encourage domestic energy production. Oil production can be promoted by incentives such as special tax treatment and selective import controls. In particular, the development of Alaskan oil can be encouraged. Funds can be allocated for the development of alternate energy forms and alternate means of producing oil, such as hydrogenation of coal. But the development of new

---

[4] See Charles Issawi, pp. 111-22, for a more detailed discussion of oil and Middle East politics.
[5] *Oil and Gas Imports Issues*, p. 157.

forms of energy seems unlikely in the short term.

Domestic solutions will not solve the problem of dependency. Therefore, the United States must seek ways to alleviate the national-security
consequences of foreign dependence. Several measures are possible.
The United States can diversify foreign sources. The most likely alternate sources are Canada and Venezuela. A diplomatic agreement
with Canada for the free importation of oil and a common Canadian-
American energy policy have been proposed. The United States has
also discussed a treaty with Venezuela to permit the development of
heavy Venezuelan oils and to provide for free entry of these oils into
the United States in return for investment guarantees to companies
developing the oils. These Western sources, however, will be unable
to fill the energy gap in the near future.

In case of embargo or supply interruption, the United States can
seek to improve its bargaining position through stockpiling. At present, the American government has no stockpiling policy. Domestic
production and industrial reserves are the only stockpiles the nation
has to meet an emergency. Stockpiles of 60, 90, or even 180 days
have been proposed. Although the high cost of maintaining such reserves and the concerns of environmentalists are two real problems
with stockpiling, there seems to be a general consensus that some
degree of stockpiling is necessary. The question is how much.

The United States can seek to improve relations with the Middle
East and North Africa. The realization of an Arab-Israeli settlement
would be the most productive, if the most difficult and unlikely, part
of such a policy. In the economic sphere, the United States can show
greater concern for the economic development, trade preferences, and
technical assistance that this region seeks, and it can establish a more
realistic policy toward nationalization and pricing goals of OPEC.
While all analysts agree that a broader base of political understanding would be helpful in the energy-security policy, many doubt the
potential for cooperation before the Arab-Israeli conflict is resolved.[6]

The United States might seek cooperation on energy policy with
Western Europe and Japan. Both are significant energy importers and
much more dependent on Middle East and North African oil than the
United States. Western Europe, which depends on imports for most

---

[6] U.S., Congress, House, Subcommittee on Foreign Economic Policy, Committee
on Foreign Affairs, *Foreign Implications of the Energy Crisis*, 92d Cong., 2d. sess.
September 21, 26, and October 3, 1972, pp. 152-72.

of its oil needs, obtains 85 percent of its oil from the Middle East and North Africa. The degree of dependency may decline for Great Britain, Norway, and the Netherlands as a result of oil discoveries in the North Sea, but the general high dependence for the rest of Western Europe will persist. Japan's situation is similar. That country has no oil, though there are hopes of some discoveries in the Sea of Japan, and it imports 90 percent of its oil from the Middle East.

Although the interests of America, Western Europe, and Japan are not identical, these countries have good reason for cooperation or at least consultation on energy policy. There is a common interest in developing alternate energy sources that can reduce the degree of dependence on oil from the Middle East and North Africa. New sources of hydrocarbons and new forms of energy are of common interest to all. Cooperation can speed development and lower costs for individual nations. A common policy in relations with the producing countries can be of interest to these consuming states. Cooperation can improve bargaining positions with OPEC and encourage the development of stable, long-term, import-export relations. Cooperation can also avoid cut-throat competition for available energy in times of shortage.

Finally, the United States should maintain its naval presence in the Middle East to deter the Soviet presence and should continue to affirm the strategic importance of this area and of transportation routes for American national security. The threat of American military reaction to any Soviet interference with Western oil supplies from the Middle East and North Africa has been an important force in preventing any Soviet action. Continued American military power and commitment to its use will be an important means of securing energy supplies.

While American dependency on foreign oil sources is inevitable in the short term, it will not be inevitable after 1985. By that time alternate sources of energy should become available: coal and its derivatives, nuclear energy, shale oil, geothermal energy, and solar energy. For the development of these new sources of energy, government financing will be required in the near future. An important contribution to the development of these sources, again, would be cooperation of the United States, Western Europe, and Japan.

While self-sufficiency in the future is possible, is it desirable as a national-security policy? Are the costs of self-sufficiency worth the price? A policy of complete self-sufficiency might require such a large share of national resources that it would lower the growth rate and standard of living and possibly damage the economic basis for national security. A policy of total dependence would be economically advantageous but would make the United States vulnerable to national-security risks. These risks, however, change at different levels of dependence. National-security planners must balance the risks against the economic consequences.

More fundamentally, it is questionable whether self-sufficiency is the way to solve the energy-security problem in the near future. Advocates of self-sufficiency may not be correct in arguing that the only way to assure adequate domestic supplies for national security is to exploit American oil fields for current consumption. Perhaps a more effective national-security policy would be to preserve American reserves in the ground. While a national policy of setting aside large domestic reserves with standby wells drilled and capped would assure the immediate availability of these reserves in time of a security crisis, a permissive import policy would minimize the use of American domestic supplies.

In any case, it seems inevitable that the United States will have to learn to live with a greater degree of energy dependence than it has known in the past. Although such dependence will require new policies, it will not result in a national-security crisis.

Energy and the Balance of Payments

In the 1970s major changes may occur in the international economic sphere. It already seems clear that the international monetary arrangements and trade patterns of the 1960s will not meet the needs of the next few years. The dollar is no longer the linchpin of the international monetary system. European and Japanese products are often as good as or better than those made by American competitors. The United States, which has been self-sufficient in energy resources, will become increasingly dependent on energy imports. It will thus become an important competitor of Western European countries and Japan, which have long needed to import energy supplies.

Traditions die hard, even after developments have made them hollow. Self-sufficiency in natural resources, especially energy resources, has been a strong tradition in the United States. Clinging to this tradition, however, could lead to misdirected policy decisions. If the United States is to maintain its standard of living, it cannot maintain a policy of self-sufficiency. It must realistically consider options within clearly defined constraints.

World Petroleum Supply and Demand in the 1970s

The overriding constraint is, of course, the availability of energy. The entry of the United States into the world market on a large scale and the rapidly growing demand in other countries may cause soaring prices, at best, or acute world shortages, at worst. It is well enough to say that it is not a question of running out of energy resources but of finding new ones. Developing energy sources, however, is a long-term process even if expensive crash programs are adopted. As Lord Keynes once said, "in the long run we all will be dead." The energy crisis, if it exists, is a short-term shortage of petroleum and its derivatives.

The short-term prospects for substitutes are not bright. Worldwide, little if any immediate relief can be expected from tar sands, shale, nuclear power, geothermal energy, and pipeline-quality gas from coal. Energy from tides and the sun is even less likely to provide relief. Government-funded research may hasten the day when these sources will be available. However, experts agree that these sources will not help much in this decade.

Rising from $1.6 billion in 1960 to $2.7 billion in 1970, United States petroleum imports grew more slowly than overall imports when quota restrictions were in effect. By 1972, petroleum imports had risen by 60 percent to $4.3 billion. Some observers expect that by 1980 petroleum imports will total $25 billion, if all potential demands are satisfied. That is an alarming figure for those who are concerned about the United States balance-of-payments deficit.[1] Those preoccupied with self-sufficiency find little comfort in the increase in the import component of total needs from 21 percent in 1970 to about 50 percent ten years later.

Perhaps 10 million of the 12 million barrels that will be imported daily in 1980 will come from the Middle East and Africa. Even though these areas are presently minor suppliers, they have large reserves. While the United States output is likely to remain fairly constant at 12 million barrels daily for the decade, Saudi production of 6 million barrels daily is expected to exceed the United States figure by 1976 and is targeted for 20 million barrels daily by 1980. The total is realistic; proven reserves are presently 145 billion barrels, which is roughly four times that of the United States. These reserves continue to rise. If the current price doubles by 1980, then clearly the absorptive

---

[1] Chase Manhattan Bank, "Business in Brief," no. 109, April 1973.

capacity of the Saudi Arabian economy, with close to $20 billion of oil exports a year, becomes an important issue.

Payments Deficits and the Monetary System

Cumulative deficits in the United States balance of payments have created a hoard of Euro-dollars estimated as high as $90 billion. Effective dollar devaluation since July 1971 of 31 percent against the mark, 26 percent against the yen, and 23 percent against the French franc have done little to stanch the outflow. Continuing dollar accumulations abroad plus a floating dollar rate will cause a secular downward movement in the dollar's value. It is no wonder that assaults on foreign exchanges have become more and more frequent and the dollar has become the principal target. The marksmen are generally identified as avaricious gnomes of Zurich and oil-rich sheikhs from the Middle East looking for a fast profit at the expense of others. Less noticed but having far more resources are multinational corporate treasurers who are also engaging in this speculation. In any event, the issue is not what is morally right, but rather why the world's monetary authorities have not replaced the defunct Bretton Woods Agreements with a new system that works. It might also be added that as the dollar reserves of the oil-producing countries continue to grow, if indeed they do, their interest in maintaining the value of the dollar will also grow. There is an old saying that if a debt is small, the debtor is in trouble; but if it is large, the creditor is in trouble. Considering the monetary situation, this saying should be translated into Arabic.

Before examining the relationship between the energy situation and the United States balance of payments, it would be well to review briefly events of recent years. Essentially, a nation's balance of payments reflects the values of imports, exports, and long- and short-term capital movements over a year. If inpayments are greater than outpayments, the balance is said to be in surplus. If outpayments exceed inpayments, as has been the situation of the United States almost every year for more than two decades, then the balance is said to be in deficit. In other words, export proceeds and foreign-capital inflows have not been sufficient to offset payments for imports and capital outflows. As a result, the United States has been a debtor nation.

If the United States is in deficit, then other nations must have surpluses and must be creditors of the United States. For many years, creditors had two options for settlement of the debt. They could ask the debtor to draw down his savings (the noninterest-bearing gold stock), or he could carry the deficit balance not only in interest-bearing investments in United States securities but also Euro-dollar deposits, probably in a London bank. In fact, most creditor nations did both.

Since August 1971, however, the first option has been foreclosed by the president's suspension of convertibility of dollars into gold. The "savings account" has been frozen, and the surplus nations must extend credit whether they like it or not. Even worse for the creditors are the two devaluations of the dollar, revaluations of the yen and mark, and a dollar floating downward as inflation continues in the United States. If this particular scenario represents respectable money management, then the nervous gnomes of Zurich, the oil-rich sheikhs, and the multinational corporate treasurers who move from country to country deserve equal respectability.

Energy Imports and the Balance of Payments

Although energy costs are a small part of the balance-of-payments problem, they are a vital concern in a high-energy-using industrial economy. The remainder of this paper will explore ways in which energy could affect the balance of payments during the rest of the decade.

The components of the balance of payments—i.e., imports, exports, capital movements, and deficit—provide a means for examining the problem. The deficit could be narrowed by reducing outpayments, increasing inpayments, or both. Smaller volumes of imports and capital flows abroad would reduce outpayments; more exports and capital inflows would increase inpayments.

Enough has been said here and elsewhere to show that total outlays for energy imports cannot be reduced during the 1970s. There are possibilities, however, of reducing the rate of increase, although some unfortunately are more apparent than real. There are further possibilities of increasing exports, although not all would be directly related to energy. Finally, there are opportunities to reduce capital outflows and to increase inflows.

Since costs of crude oil are a fraction of final-product prices, rising crude prices could be offset, at least in part, by lower product prices made possible by improved technological efficiency in other phases of the industry. Products with high-energy inputs could then be more competitive with foreign products in the domestic market and in the world market. Much of the technology is available. In most circumstances, for example, huge cost savings per ton-mile could be effected by increasing tanker sizes. Very large crude carriers (VLCCs), however, require deep-draft ports or offshore moorings where crude can be carried ashore either by smaller shuttle tankers or pipelines. While Western Europe and Japan already have several superports for VLCCs and are building and improving other ports, the United States has none, and few have been planned. Expansions of existing refineries and grass-roots construction until recently have been lagging similarly.

Environmental constraints and import quotas have severely inhibited technological advances in refinery expansion. One could conclude that at least part of the energy crisis is logistical. When there is a demand, suppliers find a way. To supply the lucrative United States market, superports and large-scale refineries have been built in the Canadian Maritime Provinces and in the Caribbean islands. Other refinery sources as far away as Italy have also been tapped. These manufacturing costs add to the value of the petroleum imported into the United States and are reflected in higher import costs. In the case of gasoline, if the amount demanded does not decrease appreciably as prices rise, even a tariff of 52 cents per barrel (forty-two gallons) means only that it is added to the price the consumer pays at the pump. It does little or nothing to hold down imports.

In some cases, foreign construction by United States companies has resulted either in capital outflows to finance the facilities or in the reinvestment of earnings, which otherwise might have occasioned a capital inflow in the form of dividends. Thus, there is an unfavorable effect on the capital account.

Domestic Supply Possibilities

For the domestic supply, the task is not so much increasing production as maintaining current levels. To the extent that this can be done, the rising physical volume of imports can be slowed. By the end of the decade, the Alaska pipeline from Prudhoe Bay to the tidewater on the Alaskan west coast could add 2 million barrels daily to domestic output. An increase in federal leasing of sites off both coasts and off the coast of the Gulf of Mexico would hasten the exploration of many promising geological structures. In 1970 there was a three-year delay between lease sales and production in the Gulf of Mexico. If this lead time could be decreased, domestic supplies could be increased more rapidly. Increased tax incentives might encourage the development of new means for increasing the amount of oil recovered and prolonging the life of known onshore fields. They might also encourage drilling deep zones and stratigraphic traps, the two principal possibilities left on land for wildcatting. Finally, increases in the price of natural gas allowed by the Federal Power Commission, now considerably lower than the BTU equivalent of oil, would probably unlock reserves currently known but not developed.

Costs, Prices, and OPEC

Some believed that the crisis had been contrived, if not by the oil companies, then by the Organization of Petroleum Exporting Countries (OPEC). The members are Iran, Kuwait, Saudi Arabia, Iraq, Venezuela, Qatar, Libya, Indonesia, Abu Dhabi, Algeria, Nigeria, and, in associate nonvoting status, Ecuador. Although the seven Arab countries excluding Iran are big producers, some observers regard them as unstable sources of supply. One way to produce stability and lower import bills is to establish an offsetting buyers' group. This might be called the Organization of Petroleum Importing Countries (OPIC) and might operate under the aegis of the Organization for Economic Cooperation and Development (OECD), which includes the major Western European countries, Japan, and the United States. Presumably, the functions of OPIC would be both economic and political. Bargaining as a unit, it would reduce OPEC prices, while it more or less discreetly held the combined political clout of most of the world's powerful nations.

Advocates of this plan probably overrate OPEC's effectiveness. They also ignore the rent theory of Ricardian economics. To illustrate, take the following production costs per barrel as approximate: the Middle East, 10¢ and the United States, $1.31, with other major producers such as Venezuela (51¢) and Indonesia (82¢) falling in between.[2] If everybody is to remain in business, then the price must be at least $1.31, or high enough to cover the cost of United States producers, taking into account crudes of varying quality and differing freights. The difference in netback from a uniform price by United States and Middle East producers with widely different production costs is known as "rent." So long as costs differ and prices are uniform, the rent will remain.

[2] Jahangir Amuzegar, "The Oil Story: Facts, Fiction and Fair Play," Foreign Affairs 51 (July 1973), 679.

In the last few years, OPEC and the oil companies have negotiated not only the size but also the share of the rent. The proposal for an OPIC would reduce the size of the rent by lowering the price. In economic terms, the expropriated rent would become consumer surplus, defined here as the difference between the price before and after the bargaining.

With petroleum demand growing worldwide, it can be postulated that prices for OPEC oil would have risen dramatically even without the Tehran, Tripoli, and Geneva Agreements negotiated with the oil companies. The entry of the United States as a buyer in the world market has only reinforced the upward price trend. Under these circumstances, buyers are less likely to be concerned about consumer surplus than about obtaining adequate supplies at whatever price. Therefore, there is little basis for buyer solidarity.

An effective cartel or monopoly of the oil-producing countries (OPEC has been called both) must control not only price but also output in order to support the price. Rising prices are *prima facie* evidence that demand is outrunning supply. But is supply being deliberately held back by a concerted effort of OPEC members?

There are indications that suggest the contrary. Libya has cut back production drastically for a variety of economic and political reasons. Ostensibly awaiting the discovery of new reserves, Kuwait has been holding the line on production—a reasonable position from the standpoint of conservation. Iran, Saudi Arabia, and the gulf sheikhdoms, all big producers, are producing more. On a smaller scale, Nigeria and Indonesia are expanding production, and Venezuela hopes to restore its reserve position by developing its vast tar belt. Thus there does not appear to be any general agreement to hold down production even covertly. Thus rising prices can largely be attributed to burgeoning demand in the industrial countries.

Other facts also indicate that the rising prices are not a result of a conspiracy. The OPEC members are far from a homogeneous group. Their governments range from socialist (Algeria) to monarchist (Saudi Arabia) and include several variations in between. Population densities range from heavy (Indonesia) to sparse (Abu Dhabi). Their development possibilities and needs beyond the oil industry are equally disparate. Even the generic word *Arab* hides great differences. If instability in the Middle East (that is, instability in the Western sense) is a threat to the consuming countries' petroleum supply, it is not based on political mischief and capriciousness toward the West or animosity toward Israel. It should be pointed out that Arab political speeches translated into Western languages lose much of their true meaning because the subtleties of the area's politics and culture are lost in translation. Even if Israel did not exist or if a meaningful settlement with its neighbors were effected, there would remain larger problems of economic and social change, which oil wealth has partly caused and can only partly solve.

The proposal for an organization of consuming countries, then, ignores the economics of the world petroleum market and the politics of Arab OPEC members. By doing so, it could well suffer the charge itself of political mischief; worse, it could drive the producing countries into the hands of powers unfriendly to certain Middle East governments and to Western nations in general.

Just as the interests of the producing countries are diverse, so are those of the consuming nations. With the Middle East supplying almost all of Japan's oil, three-fourths of Western Europe's, and a growing proportion of the United States's, consuming countries would have a hard time formulating a common energy policy to which all would adhere. In the present atmosphere, a policy of no policy (that is, every man for himself) would result in fewer conflicts of interest than any formal arrangement with inevitable complications.

Middle East Supply, Earnings, and Absorptive Capacity

Each major consumer has its primary supplier. The United States has its domestic production, which includes Alaskan and offshore wells; Western Europe and the United Kingdom have the North Sea; and Japan has Indonesia and other smaller possibilities closer to home. During a shortfall, these consumers have to augment their supply with "swing" oil from the Persian Gulf and Africa. The alleged insecurity of these latter sources is believed to threaten the continuity of supply and the stability of Western currencies, most notably the dollar. The allegation of insecurity must be examined, if only briefly, country by country.

Starting from the west, the first major producer is Algeria, which has large reserves of crude and huge reserves of natural gas. With a socialist ideology and a history of nationalization, the country has apparently settled down and is ready to make long-term arrangements. Algeria also has a large population and ambitious development plans.

It is in the interest of all parties to honor the several existing long-term contracts for crude supply. The West needs the oil, and Algeria needs the foreign exchange. Of equal interest, considering both supply and balance of payments, are the arrangements that have been made for liquefied-natural-gas (LNG) exports. Pipelines, liquefaction plants, and LNG tankers are highly sophisticated and costly equipment. Both American industry and the export accounts in the United States balance of payments will benefit from the over $400 million credit extended to Algeria for these items by the Export-Import Bank and a consortium of commercial banks. According to the terms of the agreement, Algeria will repay the loans over twenty-five years while it supplies the liquefied natural gas critically needed in the United States. Evidently, differing ideologies are no barriers to agreement, if the terms are right.[3]

Libya's revolutionary regime seeks an identity and a role. Complete takeover of the fields by the government seems the most likely prospect; however, once this takeover has been accomplished, the bases for new and durable supply arrangements may be established, as in Algeria. In the Libyan case, patience by the consuming nations may not only be the better part of discretion at this stage but a positive virtue.

Because the really large amounts of "swing" oil must come from Iran and Saudi Arabia, their impact on the balance of payments requires particular attention. Iran has adopted a 1980 target of 8 million barrels daily, roughly double the production at the beginning of the decade. The government has big economic and political ambitions. Development plans, which require importing capital goods, will absorb large quantities of foreign exchange earned from oil. The United States export accounts will benefit to the extent that American businessmen can compete successfully for the business.

Politically, Iran sees itself as filling a power vacuum created by the British military withdrawal from the gulf and the growing Soviet influence over its neighbor Iraq; it has also been long concerned about its common border with the Soviet Union. Therefore, Iran is in the market for the most advanced and sophisticated United States weapons systems. Delivery of this equipment would not only provide an immediate boost to the export account but also the continuing income from spare-parts shipments and technical assistance.

Iranian investments overseas may be a harbinger of the future

[3] The same is true of the arrangements for LNG being negotiated with the Soviet Union by several United States companies and Japan. It is unlikely, however, that the planned facilities will begin operation in this decade.

activities of producing countries. Iran's participation in oil refineries in India and South Africa and its joint ventures in North Sea explorations are examples. Perhaps most interesting of all is an agreement between the National Iranian Oil Company and a United States concern whereby the former guarantees a long-term supply of crude oil in exchange for a share of the domestic refining and marketing assets of the latter.[4] In effect, outpayments for oil over the period would be offset in the balance of payments by inpayments in the form of investments in facilities.

Among the big producers is Saudi Arabia, whose ability to absorb large inflows of foreign exchange is questionable. Pumping 20 million barrels daily by 1980, it will be the world's largest producer even though it has a sparse population and few natural resources other than oil. Although some Western observers may question the determination of the government to buy modern weapons, this policy has its own logic in a regional context; two of its neighbors, Oman and South Yemen, are receiving military aid from the Eastern bloc. The experience of the government's Petroleum and Minerals Organization (Petromin) suggests that the absorptive capacity of foreign exchange may be larger than it appears at first. Petromin has been the moving force in establishing many successful industrial enterprises, and in planning many more. Oversubscription of stock offerings to the private sector indicates that the Saudi investor is interested in more than gold, Swiss bank accounts, and real estate. One real constraint on the country's economic growth at present, however, is the deficiency of trained manpower, which the country is rapidly correcting. Hundreds of Saudis are studying technical subjects at the government's Petroleum College and abroad.

Charles de Gaulle once said that the United States's balance-of-payments deficit was financing its takeover of European industry. He reasoned that chronic deficits created a huge Euro-dollar market in which American corporations borrowed funds to buy up continental companies. Thus, American fiscal irresponsibility tipped the ownership pattern of companies in Europe toward United States multinational concerns. The United States, however, assured Europe that there was no possibility of a significant American "corner" on European industry. The industrial establishment was simply too large and growing too rapidly for limited United States resources to master it.

[4] *The Wall Street Journal*, July 27, 1973.

Recent trends of foreign investment in the United States suggest that it is the Americans' turn to be reassured. Foreign investments, especially by European and Japanese firms, are now being used for investments in plants and equipment in the United States. Although the inpayments represented by capital inflows are a favorable development in the balance of payments, fears of a foreign takeover are being expressed in both government and private circles. If the Europeans and Japanese are already here, can the Middle Easterners be far behind?

To answer, certain facts must be brought into focus. In order to sustain a growing economy whose annual gross national product is already in excess of a trillion dollars, it has been estimated that capital requirements for petroleum production and distribution facilities alone must average around $30 billion a year, which is almost double the present level. These requirements must be considered along with current investment in all United States manufacturing industry plus utilities of around $42 billion a year.[5] The conclusion is inescapable that the United States petroleum industry needs all the financial backing it can get from whatever source, domestic or foreign. Even if the wildest estimate of free reserves held by the producing nations (Saudi Arabia in particular) proved accurate, those reserves would not even finance the annual incremental investments needed by the United States petroleum industry alone.

Another conclusion appears inescapable. The United States position of world economic supremacy throughout most of the 1960s has given way to a position of *primus inter pares* in the 1970s. Once that fact is recognized, the flow of investment, technology, ideas, and people from abroad, actually a repetition of the country's economic history of the nineteenth century, should be regarded as a definite addition to the national patrimony rather than ominous threats to sovereignty. Only half facetiously is it noted that there are no Saudi marines.

One editorial writer has stated the case very well: "When the rest of the world has as much invested in the United States as the United States has invested in the rest of the world, mutual security will rest not so much on treaties or a symbolic military presence as on the clear perception that everyone is in the same boat."[6]

---

[5] National Petroleum Council, *U.S. Energy Outlook*, vol. 1, July 15, 1971.
[6] *The Wall Street Journal*, June 22, 1973.

The Nature of the
Petroleum Industry

WINSTON W. RIDDICK

The three major components of the oil industry are exploration, production and refining, and distribution. The purpose of this paper is to examine these components of the oil and natural-gas industry in the United States, with some references to the Middle East.

For about ten years while the United States's reserves of oil and natural gas have been declining, world production has been sharply increasing, especially in the Middle East. World crude-oil production doubled once between 1945 and 1950 and again in the 1950s. Between 1960 and 1968, world crude-oil production again doubled to a total of 2 billion tons.

Since World War II, United States consumption of natural gas and oil has increased rapidly. Because production has increased less rapidly than consumption, the reserves of natural gas and oil in the United States have declined markedly. For example, in 1972 the United States produced more than 3.2 billion barrels of oil and added only another 1.5 billion barrels to its reserves. The net result was a decline of 1.7 billion barrels in the oil reserves of the United States, which then totaled 36.3 billion barrels. At the present rate of consumption, the United States has approximately eleven years of oil reserves. The same pattern marked natural-gas consumption, production, and reserves. Although an additional 9.8 trillion cubic feet of natural-gas reserves were discovered in 1972, 22.5 trillion cubic feet were consumed so that proven natural-gas reserves declined by 12.7 trillion cubic feet. The gas reserves in the United States now total 266 trillion cubic feet, which is the equivalent of approximately twelve years of reserves at the present rate of consumption.

This decline in both oil and natural-gas reserves is not a new phenomenon in the United States. After peaking in the early 1960s, oil reserves began a decline which was briefly interrupted by the discovery of oil in northern Alaska in 1968 and then continued even more rapidly than before. Peaking around 1967, natural-gas reserves have declined ever since.

The increasing consumption and the declining quantity of proven reserves necessitated more importing of both oil and natural gas in the 1960s. This trend will continue at an increased rate in the 1970s and 1980s, according to most government and petroleum-industry experts.

Exploration

In the United States and the Middle East, exploration for oil and natural gas is influenced by geological considerations, the right to explore, and cost. Oil is said to be found in pools, but in reality it is found in porous rock and sand beneath the earth's surface. The geological formations and structures determine the type and quantity of oil and the best means for discovering and producing it. Oil is brought to the earth's surface by either a natural-gas drive or a water drive. That is, natural gas may trap the oil, making it rise, and pushing it to the surface; or rising water may bring it to the surface. Natural-gas drives are most common in the United States and water drives in the Middle East.

Usually, oil is found in sand formations in the United States and in limestone formations in most of the Middle East. Limestone formations provide fewer difficulties in production because limestone does not clog up the well. Sand formations, on the other hand, require a slower rate of production and very careful controls to keep the well from "sanding in." Once a well "sands up," production can be resumed only after redrilling removes the sand from the well.

The Middle East has both superior geological formations and structures and a water-drive system that is capable of producing greater quantities of oil with fewer problems of conservation. Because its oil and gas are nearer the surface, Middle East wells do not have to be drilled to as great a depth as United States wells.

In the early days of oil exploration in the United States, oil men operated on a "law of capture," which provided that whoever found oil owned it. Once oil was discovered on one piece of property, neighboring property owners drilled offset wells to tap the same pool. As a result, more wells were drilled than were necessary for the efficient production and conservation of the oil and natural gas. In the late 1930s, states legislated two major limitations on the "law of capture." To obtain the right to drill an oil well, one was required to secure from a state regulatory agency a drilling permit that fixed the location and depth of the well. This regulation limited drilling to as few wells as necessary. To determine ownership of oil beneath the earth's surface and thereby to protect the property rights guaranteed in the fifth amendment, pooling of proprietary interests was used to determine the interest of each landowner and royalty owner in a specific pool of oil. This step was effective in conserving oil as well as limiting unnecessary drilling.

In the Middle East, however, the right to explore for oil rests upon an entirely different basis. In most Middle East countries exploration is not governed by mineral codes or statutory provisions but by concessionary agreements between the national governments and major oil companies. These arrangements are similar to those made by the European kings in establishing crown trading companies for the colonization of the new world. The concessionary agreements provide for exploration, the division of oil royalties, and all other matters relating to the exploration, production, and marketing of oil. These documents are brief and precise in stating the interests of the national government of the country in which oil exploration is undertaken. Drilling permits and spacing of wells, however, are not a major consideration in the Middle East because the limestone and water drive generally make conservation a less important factor than in the United States. Nevertheless, oil companies have generally regulated their own exploratory efforts because the concessions are usually large tracts of land which permit them to develop the oil fields without any dangers of competitive lessors' and adjacent property owners' attempting to tap their oil pool by drilling offset wells.

The costs of exploring for oil and natural gas are astronomical. A 15,000-foot well drilled in the United States onshore may cost as much as $1 million. A 15,000- to 20,000-foot well drilled offshore on the continental shelf sometimes costs $1.5 million. With such a high cost for exploration, it is necessary to have a substantial return on an investment. Oil men call the petroleum industry a business for "high rollers." During recent years, however, United States oil and gas exploration has been caught in a tremendous price squeeze. For example, the average cost of drilling wells in 1971 was 85 percent higher than in 1956. During that same period, however, natural-gas prices increased only 37 percent and oil prices only 17 percent. The result of this major cost-price squeeze was a substantial decline in drilling activity. By 1971, the number of wells drilled annually had dropped to less than half the total number in 1956.

In the Middle East, drilling generally costs more per well than in the United States. However, fewer offset wells are drilled; multiple-completion wells are more frequent; and the wells are not as deep as those in the United States. Moreover, because of favorable geological formations and structures in the Middle East, a greater amount of oil is produced, and total drilling costs are thus lower or competitive with costs in the United States.

Exploration in the United States and the Middle East, therefore, is conditioned by geological factors, the rights of exploration, and costs. The net effect of the interrelationship of these factors on United States and Middle East oil production has been to give the Middle East a substantial advantage over the United States in increased exploration.

Production

The unregulated production of oil and natural gas often has undesirable consequences. First, it may cause price instability and gasoline price wars, as they are called in the United States. Oil, until it is needed for consumption, is best stored in its natural habitat beneath the surface, because of major surface problems of evaporation and chemical changes. Second, an unregulated flow of either oil or natural gas frequently results in vast quantities of petroleum remaining beneath the earth's surface, which is undesirable from both a conservation and investment viewpoint. Third, too rapid a rate of production of wells in the United States results in sanding up and necessitates additional expenses in reworking the well. As noted, limestone formations in the Middle East prevent this problem.

Governments limit petroleum production by regulating the amount extracted from the earth, the amount imported, and the price of natural gas at the wellhead. One major goal of the oil-conservation effort in the United States was to limit the rate of production of wells. This rate frequently determines the quantity of oil produced from a pool. Natural gas is the primary vehicle for bringing oil to the surface in United States wells. If natural gas is allowed to rise to the surface too rapidly, it does not lift as much oil and leaves vast quantities beneath the earth. Although not as great a problem in the Middle East because of the water-flow drive, too-rapid production becomes a factor in securing the highest recovery possible in an oil pool.

The major means of regulating production is to allow each well an amount that may be produced each day to ensure the greatest amount of recovery. Determining "allowables" requires considerable study of the geological factors, the depth of the well, and the nature of the oil and gas being produced. For many years, United States wells operated below the oil allowable, but during the 1960s United States wells were run at full allowables because of increased domestic consumption and restrictions on oil imports.

In the Middle East, oil allowables were not given serious considera-tion until the late 1960s. Then they were used primarily as a means of controlling the price of oil by limiting the quantity available on the world market. Conservation was not the primary motivation. Oil allowables in the United States are a conservation measure, but in the Middle East they are a means of improving prices by withholding pro-ducts from the market.

"Oil-import quotas" have been used primarily by the United States as a vehicle for protecting the domestic oil industry. President Eisen-hower called for voluntary oil-import quotas in 1954 and 1958. In 1959, however, he had to make them mandatory. The quotas provided that domestic producers would be guaranteed 90 percent of the United States market. Several consequences for the United States and the Middle East resulted from the decision to develop oil-import quotas. First, these quotas encouraged domestic oil production and explora-tion. Some oil experts have estimated that if these quotas had not been imposed, from 33 percent to 50 percent of the domestic oil production would not have occurred, because the price advantage of Middle East oil would have discouraged domestic oil exploration and production. Many oil men think that the Alaska oil discoveries and many offshore discoveries would not have been made without the incentive of a quota, because oil in the United States could not be discovered and produced as cheaply as Middle East oil. Second, quotas have kept the price of oil at an artificially high level for United States consumers. Likewise, they aided coal and other competitive energy sources in ex-panding their efforts domestically. Third, quotas have encouraged the depletion of United States reserves at a faster rate because they provide for greater consumption of domestic oil than would have occurred if a free-market operation had been allowed and cheaper Middle East oil had been imported.

Moreover, during the period of a buyers' market, quotas allowed the United States greater influence in the pricing of Middle East oil. It was not until the Middle East producers united and domestic reserves in the United States declined substantially that the oil market shifted to a sellers' market and the quotas became meaningless, because the United States needed to increase its importation to meet domestic needs. Therefore, quotas have provided not only for price stabilization for United States domestic oil production but have also served as a restraint on Middle East oil expansion.

In the early 1950s, the United States Supreme Court held that the Federal Power Commission had the authority to regulate the price of natural gas at the wellhead, the point of production. Exercising that power, the commission has kept the price of natural gas at artificially low levels. The average wellhead price for natural gas showed only a moderate rise between 1950 and 1970, increasing from a mere 10 cents to 20 cents per thousand cubic feet. There were two major consequences of this artificially depressed price for natural gas. Exploration for natural gas and oil was discouraged by the lower price of natural gas. Since oil could be imported for a cheaper price, there was little incentive for the most important of the integrated international oil companies to explore for oil.

However, natural gas can be produced, transported, and marketed in the United States more cheaply than it can be imported. From 1956 to 1970 when there was little incentive for additional drilling, the number of wells drilled declined from 57,000 to approximately 30,000 annually. Adjustments in the wellhead price of gas in 1971 and 1972 and a different base for establishing a reasonable price for natural gas have led to increases in its price and a major acceleration of exploration. Another major consequence of the artificially depressed price has been that it accelerated industrial, commercial, and residential dependence on natural gas and thereby discouraged the development of alternative energy sources such as coal and thermal energy. It also encouraged the development of a national network of pipelines to transport natural gas. These pipelines, a major capital investment, provided a system to transport any gas discovered by additional exploration. Also, a major problem in gas production is that it must be located in areas with access to pipelines or the cost of liquefaction will price it out of the market. The Middle East, because of the nature of its oil and gas production, does not produce natural gas in the quantities that would provide for a competitive price after liquefaction and transportation costs are added to the base price.

Thus, oil production is regulated in the United States for the purposes of improved conservation which makes possible increased rates of recovery of oil and natural gas. In the Middle East, oil production is regulated in order to maintain some price stabilization. Price regulation of natural gas in the United States domestic market, on the other hand, provides a means of discouraging competition in alternative energy sources and discouraging further exploration for oil and natural gas. Increases, however, in the domestic price of natural gas have shown a substantial, beneficial effect by increasing exploration and reducing the competitive position of alternative sources of energy.

Transportation and Refining

Most of the oil that the United States purchases in the Middle East is not refined there but in United States refineries. Until the 1970s, these refineries always operated far below their full capacity. However, in the early 1970s when most of them began to operate at full capacity, a need for additional ones developed in order to keep pace with the rapidly increasing consumption rate. To be sure, refinery plants have existed in the Middle East for many years, though their development did not become extensive until the late 1960s. The refining of petrochemical products provides many by-products in addition to gasoline. The entire petrochemical industry of the United States is based largely on the ability of its refineries to use these by-products.

Natural gas must be liquefied in order to be transported from other parts of the world to the United States. Because this process is expensive at present, natural gas transported through pipelines in the United States has a competitive price advantage. Although further exploration in the United States will probably meet domestic needs for natural gas, the need for imported petroleum and its many by-products will require considerable increases in the nation's refining capabilities and probably the development of substantial refining operations in the Middle East and other oil-producing areas of the world.

The amazing thing about the vast quantity of oil produced in the Middle East is that even though it must be shipped thousands of miles to the United States, it is still cheaper than domestic oil. The transportation of petroleum has become a major concern of the integrated international oil companies, and the development of gigantic oil tankers to transport this oil clearly illustrates the long-term significance that they place on foreign oil. As imports continue to increase, it will be necessary to develop some better techniques for unloading this petroleum.

Development of superports has created major controversies. These superports must be located at points where supersize oil tankers can anchor and connect either with oil pipelines or with smaller tankers that transport the oil to shore. Construction of superports will be extremely expensive but necessary, unless the United States eliminates its dependence upon imported oil.

Another major transportation problem involves moving natural gas to its point of consumption. The liquefication of natural gas from foreign fields is more expensive than its production within the continental United States. Increases in domestic prices will accelerate the rate of drilling and will provide an incentive to continue the use of natural gas for residential, commercial, and industrial purposes. The existence of an extensive network of pipelines for natural gas mandates continued exploration in areas adjacent to these pipelines.

Prospects

Natural gas provides the most promising area for development of the petroleum industry in the United States. With only twelve years of proven reserves still in the ground, one would not expect this to be the case. However, the industry estimates that there are probably undiscovered gas reserves of 1,178 trillion cubic feet or fifty times more gas than is now marketed annually. Over 60 percent of this potential, however, lies in areas that will be hard to explore and develop. About 14 percent of it is at depths of over 15,000 feet, and 20 percent of it is in offshore waters. Some 28 percent is in Alaska. Nevertheless, this estimate of additional natural-gas reserves is reassuring to some degree. The extensive pipeline network developed throughout the United States and the cheaper cost of domestic natural gas indicate that the nation may continue to produce natural gas cheaper than it can be imported from the Middle East.

Oil, however, presents a different picture. The prospects for increased discoveries in the continental United States are rather slim. With the exception of Alaska, offshore areas, and to some extent federal lands, there is little prospect that major oil discoveries will occur in the United States. Though many old fields are being "rediscovered" by drilling to depths of 15,000 to 20,000 feet, the possibilities of extensive oil discoveries are limited.

In addition to encouraging alternative sources of energy, there are several research areas that can be developed by the federal government to encourage and develop oil and gas production in the United States. In recent years the oil industry itself has developed several secondary recovery techniques to increase the rate of recovery of oil in proven fields. One example is the fire-flow technique. There has also been discussion about using underground nuclear explosions to rearrange geological factors in such a manner that exploration is easier and production greater. Another major area of research is the conversion of tar sands and shell oil into some form of petroleum. Conversion cost, ecological considerations, and numerous other factors, however, may limit possibilities of conversion.

One inescapable conclusion is that in the foreseeable future there will be an increased dependence upon foreign oil, especially oil from the Middle East. At present and in the future, the Middle East can produce oil in greater quantities and at cheaper prices. The national demand for oil is simply growing at a faster rate than exploration and proven reserves can meet. Although price increases for foreign oil in recent years have closed the gap significantly between the price of domestic and foreign oil, the domestic-oil supply is still not sufficient to meet long-term needs at competitive prices.

In the future, there will be significant efforts in other areas of the world to discover and develop oil and gas, which will lessen American dependence on the Middle East. Major discoveries in the North Sea and Siberia and explorations in Latin America, especially in Ecuador, Peru, Bolivia, and Brazil, all indicate that major efforts will be made to increase the amount of oil and natural gas available to the United States from other foreign markets.

Oil exploration, production, refining, and transportation in the United States have taken place in an efficient manner with the greatest emphasis on conservation of precious natural resources. By simple, geological blessings, the Middle East happens to have a competitive advantage over the United States. Questions of land tenure, due process, and exploration rights have little effect on who will have the greatest advantage in the production, development, and exploration for oil and gas.

Energy Policy and Politics

Federal Regulation
of Energy Production

GENE P. MORRELL

The effect of federal regulatory agencies on the production of energy may well be one of the more important and decisive aspects of the energy problem in the United States for years to come. Obviously, the antithesis of regulative constraint is incentive. But neither complete regulation nor the complete lack of it will solve the problems. Some moderation in regulation and a reasonable increase in domestic incentives would be the best policy.

The New York Academy of Sciences recently published an article entitled "The Future Will Be: One, Absolutely Terrific; Two, An Unmitigated Disaster; and Three, Neither of the Above. Check One." The future, the article asserts, is not what it used to be.

The entry of the United States into the world market created a domestic situation different from one that occurs when a nation accustomed to inadequate supplies enters the market. The United States is geared to surplus, not to scarcity. As recently as the 1960s, it had a surplus of energy supplies. While it consumes one-third of the world's energy, it now must compete with Japan, Western Europe, the USSR, China, India, and many other countries for world energy supplies. The regulatory climate now prevalent in the United States is not responsive to the full energy needs of the nation. Federal, state, and local governments must plan to deal with changing world conditions.

The need for energy supplies and a healthy balance of payments requires the development of adequate domestic supplies of clean fuel at reasonable prices in as short a time as possible. This simple idea entails innumerable political, environmental, technical, and regulatory problems, both domestically and internationally.

On the world market there can be no regulations that apply equally to both the United States's resource industries and their suppliers overseas. In a seller's market the domestic policies of no one country or group of consumers can have a great impact.

The United States, however, has enough domestic resources to solve its own problems. Time and technology could restore the self-sufficiency lost in 1967, if the nation had a unified policy. There are, however, some sixty-one federal agencies that deal with energy problems and regulations. In addition, each of the fifty states has its regulatory bodies, and innumerable municipalities and other legal jurisdictions attempt to deal with energy problems. These agencies must recognize that they are all under one tent and that the price of admission is controlled by the Arab world. The nation does not have time to pursue 111 or more objectives if it is to buy back its tent some day. The United States is far behind in developing its own resources. It will take at least ten years to catch up. Brandishing fists and raising voices to blame the practices that brought about the situation merely wastes time.

To understand the difficulties involved in regulating the production and use of energy, it is important to analyze resource development, the long process by which an idea of a geologist results in the availability of energy to a housewife at the turn of a switch. Between these two elements lies the basis for all federal, state, and municipal regulations. Also, between these two elements are interposed those objectives for which most Americans aspire, including cheap fuels extracted and transported in an environmentally sound manner and consumed in an environmentally sound way. Other factors influence resources development. National security demands a secure supply of energy, and there are far-reaching international implications in assuring such a supply.

Since regulation is a problem concerning people, mediocracy cannot be the common denominator of government regulators. Regulatory agencies must have technicians qualified to handle the varying problems resulting from the fuel mix that will exist by the 1980s. There is no time to cry out against industry-oriented regulation. The most qualified men may well come from the resource industries.

There is no more time for economic half-truths. Natural-resource development is an uncertain science that depends on reducing uncertainties as much as possible. Regulation adds to the uncertainty. The lack of common objectives further compounds the problem.

In order to understand the changes that must come about if Americans are to meet the world's new energy problems, it is necessary to examine the generic areas in which regulation has an impact on resource industries. This paper will consider primarily oil and gas. The uranium, electricity, and coal industries, which could ultimately supply a substantial proportion of the needs of the United States, have similar problems.

This paper will also attempt to examine the most common constraints and incentives affecting resource industries, including government regulative factors such as supply security, leasing, taxation, research and technology, transportation, environmental concerns, price, and demand. The primary objective is to determine which additional incentives or the removal of which constraints would best promote the return of self-sufficiency in the United States.

The first regulative factor is supply security, including oil and gas imports. Because of the predominance of oil in the nation's energy mix, the Oil Import Program, for all of its problems, mistakes, and alleged inequities, supports almost every other aspect of energy production. The close relationship of oil to the production and consumption of gas and the fact that the refining of oil is a relatively clean process make it a key factor in the domestic balance of supply and demand. Oil is also a great security risk in that substantial quantities must be imported to meet daily needs.

There were two reasons for the Oil Import Program. The first was refinery capacity. Because of the need for low-cost residual fuel oil, particularly on the East Coast, and the fact that supplies could be imported more cheaply than crude oil could be produced at home, the Oil Import Program allowed residual free access. As a result, refineries began moving offshore or to foreign countries. The problem was compounded in the early 1970s by an expanded flow of imported No. 2 fuel oil. The nation then exported a bit more of No. 2 refining capacity. In 1973 there is a new danger of a gasoline and natural-gas shortage.

The answer seems to be the importation of either gasoline as needed or naphtha to make feedstock for synthetic natural gas—an additional reason not to build refineries in this country. The nation must strive to keep and expand its remaining refinery capacity.

There are several reasons why domestic refineries are preferable to offshore or foreign refineries. If a nation depends only on the importation of crude oil, which it can refine to those products it finds necessary, problems of security and transportation are reduced. Crude oil can be carried in almost any available tanker. The location of refineries abroad also presents security problems. Both the source of crude supplies and the refineries must be secured. The tanker problem creates a triple security threat.

Prices on the world markets suggest, at any rate, that it is becoming less and less possible to find cheap supplies anywhere. The Oil Import Program should be maintained nonetheless, in case foreign products in a given crisis seem cheaper and more desirable politically. The industry can make long-term financial and construction commitments only with consistent planning and consistent objectives. Rapidly changing federal policies undermine long-term development.

There is another reason for avoiding foreign supplies that may be cheap for a short period. Without adequate planning objectives, domestic industries will not be able to develop any of the exotic possibilities envisioned for oil shale or coal gasification and liquefaction. Such accomplishments, which require somewhat higher prices, will not come about if the nation continues to rely on Middle East oil, which in turn jeopardizes the attainment of the goal of self-sufficiency within the next ten years.

The second regulative factor is leasing. More than 50 percent of the nation's future oil and gas resources are yet to be discovered on the outer continental shelf, which is controlled by the federal government. This area should be opened for drilling. If there is no place to drill, it is obviously impossible to find and produce oil and gas. Unless these land resources are made available, the nation's dependence on oil imports must necessarily continue to grow, subject to interruptions that have occurred on the average of once every three years for the past thirty years.

When offshore drilling is prohibited because the sale of leases is delayed, companies begin to move their equipment to other areas of the world. Offshore rigs cost up to $30 million and consequently are too expensive to be left idle. With the loss of each drilling rig, the nation loses some of its finding and producing capacity, a few of its scientists and technicians, and some of its energy independence.

The next regulative factor is taxation, a controlling area that Congress is discussing at length. The discussion of taxation could encompass the financial benefits or debits incurred by the United States economy for oil and gas development. If the industry has a price or tax incentive, new sources of fuels will be found provided the leasing system allows development. People often interpret tax incentives for the oil and gas industry as a windfall, but without some sort of incentives the industry could not invest the huge amounts of capital required for exploration.

The relationship between productivity and price is grossly misunderstood. The fact that a United States oil well produces 18 barrels of oil a day and a Middle East oil well produces 4,000 to 5,000 barrels a day does not necessarily mean that Middle East oil will be cheaper. World prices will continue to be a function of the Middle East definition of supply and demand. In the United States, the price of oil is a function of tax allowances. The Middle East will charge what the market will bear. Under present world circumstances, no matter how much oil the area has or how fast it can be produced, as long as there is some reserve, each barrel saved for tomorrow is more valuable than the barrel produced today.

Some economists think the huge differential in productive capacity means the Arabs will fail, but the record of the past three or four years in the Middle East and the past thirty-five years in the United States suggests otherwise. The Organization of Petroleum Exporting Countries (OPEC) cartel will probably stay together as well as those states in the United States that have been so often accused of maintaining controls on production to keep prices up. The question of domestic tax incentives is primarily one of whether energy costs should be spread across the board as a part of the United States standard of living or should be passed on to the individual consumer. Lowering financial incentives raises the price.

The oil and gas industry is now and probably will forever be risky. The industry's difficulty can perhaps best be explained to a consumer by an analogy between a gasoline station with ten pumps and the process of drilling for oil. The ten pumps are like slot machines. The consumer drives up to the station, chooses one of the ten pumps, and pays his dollar. He might get a dollar's worth of gas or he might get a windfall of two to four dollars' worth. Nine times out of ten, however, he gets nothing and only one time in forty he gets his money back. This analogy illustrates risks comparable to those that the industry takes in its attempts to discover and develop new supplies.

Research, the next regulative factor, is undertaken by government, industry, and universities. The problems of research are complicated by imperfect antitrust laws. Many scientists in this nation work for separate and competing companies. There are many technical and social goals that can only be attained in a given time span by allowing these scientists to work together. Because of antitrust laws, however, competing companies cannot pool either their knowledge or their scientific personnel. The antitrust office of the Justice Department should certainly retain its present watchdog attitude to ensure a competitive situation, but the nation would have a better chance to solve innumerable problems more quickly if cooperation in research were permitted.

Transportation entails a multitude of problems related to tankers, deep-water tank sites, ports, railroads, and pipelines. There is no need to reiterate the problems of regulatory agencies in this area nor the problems of the industry in trying to move fuel and follow the agencies' rules at the same time. Unfortunately, one must move fuel, whether crude or refined, and transportation cannot help but have an adverse effect on the environment. This fact must be accepted, and regulations must be devised to handle the problem.

Environmental concerns and the implementation of the National Environmental Policy Act of 1969 have caused great problems for the industry. They have brought about an almost complete stop to the development of new transportation, refining, and conversion facilities. Regulatory policies designed to preserve the environment did not cause the present fuel shortage; but, developed at a time when a real shortage was approaching, they did hasten the crisis and made it more immediately apparent. Continuation of this trend could lead to an almost total dependence on foreign oil supplies. Future regulations designed with concern for the environment must be weighed in light of the objectives set for energy resources.

The preservation of the environment need not be incompatible with the development and use of energy. Energy can rebuild inner cities, treat sewage, and perform many other tasks that will improve man's surroundings. Moreover, problems created by almost two hundred years of the Industrial Revolution cannot be solved in two or three years. A more reasonable time span is necessary to give both industry and technology time to balance their objectives for total resource development with society's legitimate demands for protection from industrial and technological pollution.

Several environmental fictions, however, must be laid to rest. A recent survey by the U. S. Coast Guard indicates that 2 percent of the world's total ocean pollution is caused by offshore drilling. Some of the media might make one think that it is the only cause. Tankers, other ships, and spillage of crankcase oil into rivers account for 98 percent of the pollution of the ocean by oil. Ironically, the objection to offshore exploration leads to a much more dangerous form of pollution from increased tanker traffic.

The relationship of prices to taxation and the international oil-supply situation, which has already been mentioned, is only one of several significant problems in this area. Congress is now considering legislation that would decontrol the price of gas from new wells so that it can reach a market clearing level. Present utility regulations, which are designed for fixed-cost installations, can never be designed to cover the risk a producer of oil and gas must undertake. Such undesirable laws place a forced demand on a commodity that is grossly underpriced in relation to competing fuels. These policies also hinder research for coal gasification, hinder the import of liquified natural gas and crude oil or naphtha, which can be gasified.

Moreover, the average domestic producer is allowed a wellhead price that is only slightly more than half of what the average foreign producer receives. Unfortunately, the legislation for decontrol before Congress is not being favorably received.

Some have suggested a unique and rather thoughtless plan whereby the intrastate market for natural gas would be controlled by the same federal mechanism that has led to the present state of affairs. Superficially, the plan seems to be an ideal way to increase supplies without an inflationary price rise by authorizing the Federal Power Commission to control intrastate gas. Although such a plan might increase supplies minimally in some areas, it merely extends the problem to other areas and compounds it. It eliminates a shortage by creating another one somewhere else in the country. If gas is taken away, it must be replaced by some other fuel. That fuel, at this time, depending on the environmental constraints, would most probably be No. 2 fuel oil. This means that an entirely new group of electric utilities and industries would have to go to the world market to obtain enough No. 2 fuel oil to keep their generators going. The increased demand on the international market for No. 2 fuel would drive the price up. As a result, the average consumer of No. 2 fuel oil on the East Coast would pay a higher price in order to subsidize a minimal supply of gas for the interstate market.

The answer to gas supply and demand is an adequate price, not substitution from the intrastate market. World demand for No. 2 fuel oil means the increase in price is beyond the control of the president, the Congress, or the Federal Power Commission. The government must adjust its thinking to that fact.

Industry, by asking people to conserve gasoline, fuel oil, electricity, and gas, has shown that it takes the problem of demand seriously. The problem of meeting the nation's enormous demand for oil, however, is also related to another crucial problem, the placement of new nuclear facilities, oil-burning electric facilities, oil-storage tanks, refinery complexes, and supertanker ports. All of these raise serious problems because no one wants them near their home, but all must be recognized as necessary to accommodate the growing energy needs of the nation.

If, as an experiment, fuel was rationed according to the refinery and producing capacity of each state, people would realize what some states are doing to meet the needs of the people in other states. Obviously, not every state would have its own refining system or producing capacity. Markets and population would not justify such a situation any more than nature would. There are, for example, areas off the coast of California, Florida, and the Eastern Seaboard that are geologically favorable to offshore drilling. To accommodate the nation's future needs, those areas must be allowed to yield their potential. At the same time, pipelines and tankers to transport oil for refining or conversion capacity are necessary. All of these items have to be accommodated in order to supply the American consumer with energy.

The last factor to be considered, organization, is of prime importance if the nation is to solve the problems discussed in this paper. There are now some sixty-one federal agencies that to some degree investigate, regulate, control, or constrain the energy industries. It would be difficult to find even one concrete, common objective among them. Each agency has its own constituency and special interest. The Department of the Interior is accused of being oriented toward resources at its best and industry at its worst. The Environmental Protection Agency is said to be overly concerned with the environment. The Department of State is said to consider foreign affairs more important than domestic problems.

The Atomic Energy Commission and the Federal Power Commission, both in their regulatory and data-gathering functions, have also been accused of constituent interests. For the most part, these criticisms are unfair and unfounded. Many of these critics have never worked in the executive branch of the government or in any of the regulatory agencies.

The president has on at least two occasions attemped to reorganize the energy area of the federal government on the basis of studies done by highly competent and technical personnel with access to information from all the federal agencies involved. He has not been successful. Congress does not want to give up its various institutional controls over the old-line agencies, a position that is aggravated by the seniority system. In July 1973 the president appointed an energy-policy director. Whether or not one agrees with the "energy czar" approach, it is one option that should be tried.

There are three agencies in the federal government that are highly competent. The Department of the Interior controls leasing and regulates oil imports. The Atomic Energy Commission regulates the use of nuclear power. The Federal Power Commission regulates the nation's utilities. These three agencies should form the core of any reorganization of the energy-regulation apparatus of the federal government. The Department of the Interior has the largest base of expertise for natural-resource research and development and thus should head the troika. Other agencies in the federal government that have a semblance of control should be placed within the jurisdiction of the present Department of the Interior, as contemplated by the president in his most recent plan for reorganization. Advisers to this department should be the Department of State, because of its worldwide influence; the Department of the Treasury, because of taxation and the obvious budgetary implications of offshore-leasing bonuses and the fee system for oil imports; and the Environmental Protection Agency, which must come to grips with both the nation's energy problem and its own particular problem of assuring a clean environment.

Americans could perhaps learn a great deal from the refreshing example of Norway, an old nation that is responding to new circumstances by developing the resources of its outer continental shelf. Norway has a unique administrative organization that allows two distinct means of communicating problems and resolving issues. One channel is the government working through an agency similar to the Department of the Interior, and the other is the industry group, an association similar to the American Petroleum Institute or the National Association of Manufacturers. Solutions to problems are sought either by the government or the industry group. There are no cries of foul play and few questions of credibility or truthfulness. Technicians solve the technical problems and politicians solve the political problems, a rather simple formula that so far has worked well for the Norwegians.

However, the United States ought not to borrow ideas indiscriminately. In Norway, for example, the oil company is owned by the state, an arrangement that cannot be recommended for the United States. The birth of the world's oil and gas industry and its technology was occasioned by the birth of American free enterprise. The formation of a state oil company in competition with such private expertise would be a mistake with grave implications. Such an operation would make the depletion allowance look insignificant. The cost to the American consumer would be more than substantial. Nevertheless, the cooperation between industry and government in Norway and the responsiveness with which they seem to be solving their problems ought to teach Americans a lesson.

The impact of the international energy situation on the United States economy and balance of payments means that Americans must reconsider the simplistic, unattainable standards to which they have become accustomed. The United States is in fact dependent on supplies from other nations. It cannot control those supplies or their price. Regulation must allow the nation to achieve its objective of self-sufficiency in as short a time as possible. The nation must look more towards the market-pricing concept as it applies to both the United States and the world. The constraints that now apply to scientists in industry must be removed in order to speed up research. Environmental restrictions must be eased so that the demand on foreign supplies does not become too great. The public must be taught the need for conservation efforts, and new ways to conserve fuel must be developed. New refineries, supertanker ports, and new coal mines are essential, as are water supply projects to the new energy basket of the nation, the West and Middle West. The industry must have access to railroads and pipelines.

It will take time to buy back the nation's freedom from the Arabs. Industry cannot meet the challenge to provide needed energy if it is hampered by restrictions that are frivolous and based on inaccurate assumptions. At present, the United States is like a jet plane. It is flying with its fuel tanks one-fourth empty. The American consumer on board has aspirations for a cleaner environment, for rebuilding inner cities.

for a vacation cabin, two cars, and a modern home with all the conveniences his nation's high productivity has made available. He has been sitting comfortably with these aspirations, but now that the fuel gauge indicates the plane may not have enough fuel to reach its destination, his reaction so far has been to stamp his feet and say that the gauge does not work or that the whole thing is a hoax. This is a crude analogy, but the American consumer has to make an informed decision. If he does not, the plane simply cannot continue. It could crash. The consumer can pay a little more, or he can stake his journey on the vagaries of foreign, primarily Middle East, politics.

Energy and Environmental Quality

ROBERT P. OUELLETTE

Although there will be no physical shortage of energy for at least a decade, there are a number of problems that have already reached crisis proportions. Repeated electric-power failures raise questions of reliability and insufficient capacity. Government regulations designed to preserve the environment may soon lead to a sulfur crisis. The most severe problem, however, is the disproportionate use of energy resources. While the supply of some resources is being diminished, others still in abundant supply are hardly being used. Although coal resources account for some 88 percent of the world fossil-fuel reserve, they will be used for only 22 percent of the national energy needs by the end of the century. Because the United States has 35 percent of the world coal reserve, this imbalance is particularly dramatic.

Although the long-term outlook for energy resources in the United States and other highly industrialized societies is excellent, with the promise of solar power, fast breeder reactors, laser fusion, geothermal energy, kerogen from oil shale, and the like, effective energy generation from such sources may be decades from realization. In the meantime, the United States remains in the age of fossil fuels. Domestic oil and natural-gas reserves are diminishing rapidly, but coal, the nation's most abundant fuel, would last for several centuries at current consumption rates.

The Sulfur Crisis

The most important factor in the fuel situation of the 1970s will be environmental regulations. In April 1971, the federal government set primary ambient air-quality standards to safeguard public health and secondary ones to safeguard public welfare for six air pollutants— sulfur dioxide, total suspended particulates, hydrocarbons, nitrogen dioxide, carbon monoxide, and total oxidants. The effective target date for primary standards is 1975; for secondary standards, a reasonable time thereafter. In contrast to the fixed standards, emission limitations on stationary sources have been set according to State Implementation Plans (SIPs), which vary from state to state. State Implementation Plans were to be submitted by January 30, 1972, to the Environmental Protection Administration (EPA), which had to approve or disapprove each portion of its plans by May 30, 1972, or develop acceptable plans for those states that failed to present plans that would meet the provisions of the law by July 30, 1972. Primary standards must be achieved within three years of the plan's approval, but EPA may grant a two-year extension, if it is requested and justified by the state. A further one-year extension may be granted for specific pollution sources or classes of sources. SIPs for secondary standards may be submitted under specified conditions up to eighteen months after the May 30, 1972, deadline and must specify a reasonable time period for compliance. Primary standards must be attained within the 1975-77 time period, although specific sources could be exempt until 1978.

Fossil-fuel combustion emits three pollutants affecting ambient air-quality standards. The proportion by weight of stationary-source fuel combustion to emission of these pollutants in 1967 was 42 percent for particulates, 43 percent for nitrogen oxides (NO_x), and 73 percent for sulfur oxides (SO_x).

The technology for control of particulates is well advanced, and control systems provide high collection efficiencies with current fuels. Thus the particulates standards should have little effect on fuel requirements.

Nitrogen-oxide control technology is less advanced, and control systems and techniques are not yet available that can be applied to all existing stationary sources to reduce emissions to satisfactory levels. For new installations, however, coal, oil, and gas-fired boilers that will meet emission-control standards for NO_x are available. It is anticipated that the initial strategy for NO_x control will be based primarily on transportation controls (transportation accounts for about 47 percent of NO_x emissions) and reductions resulting from enforcement of new-source standards.

The control of sulfur-oxide (SO_x) emissions from fuel combustion at stationary sources is much more critical since about 75 percent of the national emissions of SO_x are from these sources. Therefore, a large part of any reduction in SO_x must be achieved through control of fuel combustion. The control strategy specified by the states is to limit the allowable sulfur content for each fuel type, consistent with attainment of ambient air-quality standards. Alternatively, control systems, primarily flue-gas desulfurization, can be used with high-sulfur fuels in large boilers to attain the same objective. However, such systems will not be available for several years.

The primary means of meeting regulations, therefore, is a shift to the low-sulfur fuels. This approach will have a severe impact on the future use of coal. The sulfur content of coal types used in 1970 by the utilities industry, the nation's largest coal consumer, averaged 2.5 percent by weight. The maximum content specified by most states is less than 1 percent. Based on the assumption that present coal users will not switch to other fuels, the total coal used by stationary sources in 1975 will probably be 592 million tons. The mismatch of availability and requirements due to sulfur regulations is projected to create a potential shortage of over 300 million tons of low-sulfur coal in 1975.

Emission standards for new stationary sources have also been set for various pollutants as they apply to some industries. Implementation of these standards places a new strain on United States energy reserves by necessitating the use of high-quality, low-polluting fuels, which are in shorter supply than those widely used in the 1960s. At the same time, technology in fuel processing or postcombustion-emission cleaning is not yet reliable enough to solve environmental protection demands. The result is a crisis in the availability of high-quality fuel.

Coal, the Black Magic

Coal is the most abundant fossil-fuel resource and potentially the most damaging to the environment. It is the only fossil fuel that could, by itself, meet the cumulative United States energy demands beyond this century.

Coal is not a homogeneous material; it must further be defined as to rank (bituminous, lignite, and anthracite) and as to sulfur content. Most coal in the United States is of the bituminous and subbituminous rank, and a major portion is low in sulfur. Unfortunately, most of this soft, low-sulfur coal is located west of the Mississippi River in the Rocky Mountain area, where transportation is a major cost factor. This coal is most easily amenable to strip mining, which is currently under strong attack by environmentalists.

A shortage of 250 to 350 million tons of coal at all sulfur levels will be created if state sulfur regulations are completely implemented by 1975. Several options at reasonably low costs to increase the supply of clean coal would lower the deficit to a maximum of 150 million tons. However, the gap remains large and must presumably be satisfied by the use of sulfur-control devices or fuel substitution.

Of the technical and policy options available, only four appear to provide a sufficient quantity of clean coal at a reasonable cost in the necessary time. These possiblities are opening new mines, accelerating mining, mechanical desulfurization combined with large-scale extraction, and transportation of Rocky Mountain low-sulfur coal.

Increased Mining and Society

The apparent reason for the declining use of coal in the United States has been the undesirable side effects apparent in all phases of coal production. These effects include marginal safety in some mining conditions, unhealthy working conditions for miners, insufficient efforts to restore land spoliation or correct water-pollution practices, and failure to give adequate attention to environmental and health problems produced by both particulate and gaseous effluent releases in coal-use facilities.

Because the hazardous health and safety conditions for coal miners had decreased the rate of extraction and use of coal, the Federal Coal Mine Health and Safety Act of 1969 was adopted. Recent research by the United States Bureau of Mines indicates that health and safety problems in the coal industry can be solved at a cost of a few cents per ton produced. Although basic procedures and equipment have been developed, they must be monitored and improved.

Land desecration by the coal-mining industry is a major deterrent to widespread and expanded use of coal. The United States Geological Survey estimated that by 1970 more than 1.3 million acres had been damaged by strip mining. The National Coal Association reports that only about 81,000 acres had been redeveloped or reclaimed during 1971. Obviously, only a token effort was made. The picture is only slightly brighter for shallow- and deep-mining situations. Recently, though, some in the coal industry have recognized the seriousness of the problem and are trying to solve it. While a few reclamation projects have been quite successful, they remain the exception rather than the rule.

A recent report of the Council on Environmental Quality (CEQ) estimated that the incremental cost per ton of coal with complete reclamation using the modified block-cut technique would be seventeen cents and for contour backfilling fifty-six cents. Thus it is clear that land redevelopment or restoration would not force the price of coal beyond the range of competitiveness with other fuels.

The CEQ report also considered limiting or prohibiting surface and strip mining on slopes greater than 15 or 20 degrees. A significant amount of low-sulfur coal production would be precluded by such a slope prohibition. This loss would diminish the vitality of steel production and exports as well as the ability to meet requirements of the Clean Air Act. A combination of reclamation requirements, judiciously applied slope limitations, and a gradual shift to the use of very large underground coal reserves in the West would permit expansion in the use of coal while lessening the adverse environmental effects.

An additional objection to continued or increased use of coal is the potentially harmful emissions released at consumer sites. However, there are methods of overcoming this problem. The first is the use of low-sulfur coal, which is available, although mining it is harmful to the environment. The second is the use of desulfurized coal. The third is emission clean-up procedures.

Coal production increased by more than 50 percent between 1959 and 1969 without the impetus of a national crisis. However, a concerted effort will be required to reverse the more recent downward trend in coal production. Methods for increasing production per miner must be developed and adopted while reducing the risks and health hazards.

An increase in the use of coal will have many social effects as well as increased environmental costs. A drastic revision is needed in the education and training programs for mineral-science specialists. Technological specialists, as well as safety and environmental specialists, will be needed. The development of new coal fields will result in a significant population redistribution that will require new housing, stores, schools, roads, commerce centers, and recreational facilities.

Technological problems can be solved, given time and money, but major social and institutional barriers still inhibit a coal-intensive strategy. An all-out development of coal in the western "blue sky" and "room to roam" country will meet with major opposition. The same applies to the development of Indian lands. A large influx of miners in sparsely populated areas tends to create the social ills of migrant settlement or boom town and, after the mines are exhausted, culminates in the ghost-town syndrome.

The research and development costs and capital outlays required to put large amounts of coal into the energy system are enormous. They will be made only if there is adequate assurance of some reasonable rate of return on the investment.

Alternatives to Coal

Fuel oils—distillate and residual—can be the "swing fuels" to fill the energy gap caused by the deficit of low-sulfur coal, but their use will be limited because they are not always in the right place at the right time. It is estimated that consumption of fuel oil will rise to 1.4 billion barrels by 1975. About 2.5 billion barrels (assuming about four barrels of oil equal to the BTU content of a ton of coal) will be needed if the low-sulfur coal deficit of 250 to 350 million tons is met by switching to fuel oil that meets State Implementation Plan (SIP) sulfur-content regulations. Four options are of primary importance in fulfilling the total fuel-oil requirements for 1975: blending, increased imports, burning crude, and increased domestic desulfurization.

By blending fuel oils with various sulfur levels, about 105 million barrels of fuel with an acceptable level of sulfur can be produced at minimal cost. To obtain fuel with a 1-percent sulfur content by weight, the remaining high-sulfur oil can be desulfurized at United States refineries at an additional cost of fifty cents a barrel.

Increased imports could be used to relieve the fuel shortage. The low-sulfur coal gap of 250 to 350 million tons is roughly equivalent to 1 billion barrels of oil. About 1.2 billion barrels of distillate and residual oil could be made available on the world market from the expected 1972-75 production increases alone, a projection based on an average world crude-production growth rate of 7 percent a year. If half of this oil goes to the United States, the equivalent coal deficit would be reduced to 400 to 500 million barrels of oil. That amount could be made up by burning crude oil, rather than only the fraction of oil that is distillate and residual.

Each of the four principal options for sulfur reduction—blending, importing, burning crude, and desulfurization—have different environmental impacts. Blending will have little more impact on the environment than current operations, except for the spilling and evaporation that may result from additional handling. However, the effect on air quality of maximum blending will be an averaging of sulfur-oxide emissions from oil burning in each state at approximately the state sulfur regulation. While oil burned in each state would not exceed the state limit, its sulfur content would never be much below it.

If crude oil were imported primarily for refining into distillate and residual oils, the principal environmental impact would be at the site of new refineries. It should be recognized that these new refineries will require major investments and are likely to evolve into complex operations. Among siting considerations should be included the attitude of the various coastal states to the location of industrial facilities.

The impact of the transportation of crude and oil products will vary with the distance of the refinery from the market and the source. More imports mean more and larger tankers, which might require new deep-water ports and increase the probability of oil spills. New tankers and transfer technology could help solve this problem. If oil is to be used to fill the clean-coal energy gap, more pipelines will have to be constructed within the United States than would be needed to meet traditional requirements. The Alaska experience provides a documentation of the impact that new pipeline construction can have.

Facility siting would also be the primary environmental consideration of oil desulfurization, though processing and transportation facilities would also have to be studied. Large Caribbean refineries have a capacity of about 300,000 barrels a day. About five comparable facilities with desulfurization capabilities would have to be constructed in the United States to meet the demand. With careful site selection, this number of facilities could probably be constructed to meet environmental standards so long as these refineries did not evolve into more complex operations.

Natural gas, which now provides about one-third of the nation's total primary energy, is also available as a limited swing fuel to meet the clean-power deficit. Yet the primary means to achieving a greater natural-gas availability—increased production, coal gasification, and redistribution of supply—pose environmental problems. Increases in wellhead gas prices are expected to result in those environmental impacts associated with gas exploration and production. Onshore exploration involves well drilling and road construction. Offshore exploration generally results in the erection of structures with chronic low-level discharges and the hazards of large-scale spills. Increased gas production will also require additional processing and transportation facilities. Even though natural gas has environmental advantages as a fuel, it is a potential source of accidents because of pipeline breaks. No doubt the installation of gas-handling facilities would reduce the risk of accidents.

Arctic drilling and associated transportation activities for natural gas will have a substantial environmental impact, especially since oil will also be produced. Objections might be similar to those raised over the Alaska pipeline. A similar impact can be projected for any new natural-gas production, however small, in Mexico.

Coal gasification, an alternative means of increasing the gas supply, will also have adverse environmental effects. It will cause increased water pollution, solid-waste disposal problems, and land deterioration. Furthermore, much of the coal would come from strip mines. Thus, as the coal gasification technology is developed, an effort should be made to find ways to offset its effect on the environment. It must be recognized that these adverse effects may be much less severe than those of alternative sources of energy, such as nuclear fuels, desulfurized oil, and natural gas. The total impact of energy development for each type and location of energy source must be evaluated to determine which process has the least environmental impact.

At this state, however, liquefied natural gas (LNG) appears to have several important advantages. The environmental impact of LNG is minimal since any spills quickly evaporate. Studies indicate that the possibility of explosions, even with large spills, is extremely remote. LNG, which is discharged as a liquid, can be regasified and delivered to pipelines or directly to users. There are twelve ports on the East Coast and several on the West Coast large enough to handle standard LNG ships. Thus port facilities should not be a problem from an environmental point of view, although the possibility of accidents always exists.

A last technology to increase the supply of natural gas is nuclear stimulation—the use of underground nuclear explosions to release trapped gas. The short-term impact of nuclear stimulation on gas availability could be large only if a crash program were undertaken. However, this technology may offer significant long-range promise as a means of making new gas resources available. Conventional explosive technology and hydraulic fracturing may also provide means of unlocking these potential reserves. Environmental problems must be considered in relation to personnel and property hazards, including the impact of high-intensity explosions on ground-water supplies.

In the past it has been easier and perhaps less expensive for the consumer to turn to alternative fuel sources than to insist upon corrective measures to ensure environmental quality. The rapidly diminishing supply of cheap natural gas and domestic oil and the time scale for availability of most new energy sources have severely curtailed, if not eliminated, the option of switching to these fuels. The sole exception is atomic power, which has been widely hailed as the energy-shortage panacea.

The environmental impact of using nuclear energy will depend on the siting of future power plants, the mining of uranium ore, and fuel reprocessing. The principal concern is the emission of radioactive material. Stringent controls have been placed on these emissions not only at generating plants but throughout the fuel cycle. Under normal operations these emissions do not present an environmental hazard. The major concern is the increased chance of accident as more plants are installed. Even though the probability of a severe accident is extremely low, the accident potential may increase many-fold in the 1970-90 period as the number of nuclear plants increases from some 30 in 1975 to over 300 in 1990.

The safe storage and disposal of radioactive wastes from processed nuclear fuel is one of the most critical environmental problems. These wastes produce substantial amounts of heat for a number of years along with long-lived radioisotopes requiring hundreds of years to decay to safe levels. The quantities of these wastes will expand rapidly and their disposal will present a unique problem. Besides the high-level waste, enormous quantities of low- and intermediate-level wastes will be generated through electrical power production. It is estimated that approximately 500 to 1,000 fifty-five-gallon drums of these wastes will be shipped annually from each nuclear power facility for storage. In short, the nuclear-energy solution to the national energy problem has problems of its own.

The Energy-environment Conflict

Spurred by increasing public awareness, the justified desire for a cleaner environment has given rise to a series of environmental goals and standards. Most clean-air regulations involve the amount and quality of fuels burned. In the past, the production, selection, and use of fuels were simple questions of quantity and price. The basic equation is now made more complex by added parameters. Trade-offs must be made between quantity, quality, price, and time. If only one, two, or perhaps three of these parameters can be satisfied but not all of them simultaneously, a crisis may be said to exist.

There are four basic ways to make the best of this situation: reduce demand for energy and increase the efficiency of the energy system; increase the supply of clean fuels; adopt intermittent strategies for decreasing the environmental harm of burning high-sulfur fuels; and utilize selective pollution controls. Comprehensive reviews have been made of ways to reduce demand for energy and to achieve better economy and greater efficiency in the use of energy. Automobile horsepower and weight limits or heavy taxes on these could produce 1×10^{15} BTUs or a savings of 7 billion gallons of gasoline a year by 1985. Development of a clean, low-pollution engine, such as the stratified charge engine, could result in 1×10^{15} BTU savings a year by 1990. Installation of advanced combined (low-BTU desulfurized gas from coal gas turbine-steam cycle) power plants could mean 1×10^{14} BTU savings a year by 1985. Improved insulation in residential and commercial buildings could result in perhaps 1×10^{15} BTU savings a year by 1985. Improved appliances and lighting would produce a possible 1×10^{15} BTU savings a year by 1985. Increased use of scrap materials recycled to furnaces would result in perhaps $1\text{-}2 \times 10^{15}$ BTU savings a year by 1985. Increased price of fuels has no quantitative estimate at this time. Unfortunately, these ways are part of a long-term solution and offer little improvement for this decade.

Air quality is generally a function of pollutant emissions, the assimilative capacity of the environment, the control strategies in effect,

and ambient-background levels. The dynamic interplay of these parameters is the basis for an intermittent control strategy. Examples
of these include load switching (drawing power from outside an area
that is experiencing low-assimilative capability); work-pattern mod-
ification (staggering work schedules); and fuel switching by sulfur
quality or by type (coal to oil).

Based on meteorological conditions, criteria for evaluating intermittent control systems have been developed by EPA. Several studies
have described fuel and load-switching strategies (seasonal and daily)
to minimize use of low-sulfur fuels while still achieving ambient
standards.

The State Implementation Plans regulations were formulated by
each state authority independently, and a large number of them are
designed to go beyond the federal primary air-quality standards.
Others are not strict enough to achieve the goal. The SIP regulations
can result in huge deficits of clean coal (200 to 300 million tons a
year) and clean oil (up to 100 million barrels a year). Two reasons
for this deficit are that limited air-quality observations are used as
a yardstick and that the regulations are applied to a large area con-
taining a variety of communities and types of economic activities.

A better approach would be to tailor the control strategy to the
specified need as depicted in table 1, which shows the clean-fuel gap
rapidly closing as control is more selective and as the tailoring concept is applied to increasingly smaller units. If this process is pushed
to the utmost, the gap can be almost closed. Unfortunately, the savings
inherent in these technological solutions are rapidly offset by the
administrative costs associated with the implementation of such a
program.

A variety of tailoring approaches are feasible. The systematic use
of the crude method is based on an empirical relationship between
emission and air quality for different fuel-burning sectors.

Results of a study for three Air Quality Control regions show that
75-percent-effective control applied to a relatively small number of
emitters on a selective basis would result in achievement of the primary SO_2 standard. As shown in table 1, control of 300 emitters at
75-percent effectiveness in the New York area reduces overall emissions by 64 percent and achieves the ambient level at all receptors.
All emissions requiring control are combustion in origin. Since the
New York area may be using 5 million tons of coal and 215 million

TABLE 1

Summary of Selective Control Strategy
Applied to Three AQCRs

AQCR	Number of Emission Sources	Number of Emitters Controlled at 75% Effectiveness	Percent of Emissions Requiring Control
New York	1,285	300	64
Philadelphia	700	53	19
Niagara Frontier	541	17	27

barrels of residual fuel oil annually, less stringent emission controls on 36 percent of the fuels would provide a substantial reduction in clean-fuel deficits. These savings in clean fuel might be extended on a national basis. Furthermore, allowing variation in the degree of controls within and among source categories might assure even further savings.

Another alternative to allow more flexibility in emissions has been a proposed emissions tax. Many investigators have examined the implication of an emission tax that could provide an incentive to minimize operating costs and reduce emissions. Tax revenues could be applied toward developing better control technology and reducing environmental damages. There is hope that increased flexibility might allow most areas in the country to realize the goal of achieving environmental quality with minimal imposition of economic and social dislocation.

An analysis of a variety of energy scenarios shows that some form of selective control would provide a suboptimal solution that would be least costly and have a minimal impact on the environment. Local implementation and administrative feasibilities of such solutions have not been reviewed. Table 1 shows preliminary results of what can be done on paper. Using a systems approach, it should be possible to define the required contributions from each of the four ways acting simultaneously. The vehicle available for this implementation is the compliance schedule being developed by each state.

The 1970s are a decade of important decisions for fuel-use patterns and environmental protection. Technology will provide a price ceiling for energy and will open the door to infinite resources when coal conversions, solar energy, and nuclear power are considered. Unfortunately, these technologies will help little in the immediate future. Technology can never be the only solution; too many problems are not amenable to technical solutions.

Advances in environmental quality have been achieved after much effort, and the United States cannot declare the next ten years an environmental holiday without forever suffering the consequences. The politics of conciliation, compromise, and trade-offs is the obvious answer. The nation cannot forever expand its energy consumption without irrevocably damaging the precarious equilibrium of the American ecosphere. The extreme positions must be molded into a working-action plan. All ways must be examined. A comprehensive, coordinated research and development program must be programmed, and political and legal action must be initiated to implement President Nixon's energy message without delay.

Political Barriers to a National Policy

ROBERT S. GILMOUR

Despite mounting public pressure for government action on nationwide gasoline shortages, periodic and widespread electric brownouts or even blackouts, and shortages of heating oil and natural gas, a comprehensive national energy policy may be many years away. The finite limits of United States oil and gas reserves are genuine enough, but the reasons for this lapse in national policy, just as real, are not so much physical or even economic in character as they are political. Indeed, political factors make difficult an integrated national energy policy, which would at once guarantee an adequate energy supply while protecting existing energy and petrochemical resources. Such a policy would vigorously search out new energy sources while offering all possible environmental safeguards. No crystal ball is in hand, but a knowledge of past and present political processes and decision-making structures offers some clues to the ways in which a national energy policy might emerge in the future.

Crucial to the argument presented here is an appreciation of the fragmented character of American politics, that is, the disassociation of functionally related issues and policies. Actually, American political fragmentation is now so thoroughly taken for granted that the popular press has repeatedly reported that the "environmental crisis" of the late-1960s has more recently been "eclipsed" by the "energy crisis." This without more than the barest hint that the two might be vitally related in the world of nature. Yet obvious relationships between an Alaska pipeline through the earthquake zone and possible widespread environmental damage, between the location of nuclear power plants and thermal pollution, and between increased strip mining for coal and the resulting moonscape of land desolation hardly need elaborate explanations. The list of such relationships is a long one.

National policy-making, however, traditionally ignores interconnections such as these while the Atomic Energy Commission and the congressional Joint Committee on Atomic Energy go about the business of making atomic energy-production policy, executive mining agencies and congressional mining subcommittees proceed to write mining legislation, specialized and disparate environmental bureaus and congressional committees develop environmental policies, and so on. No attempt is made to understand the more subtle interrelationships and secondary or tertiary effects of any one of these policies. Unintended or otherwise, the effects of fragmented policies are contradictory and frustrating to the establishment of an integrated national energy and environmental policy.

To complicate matters, fifty uncoordinated state policies regulate such areas of energy and environmental concern as power-plant sites, air- and water-emission standards, public-utility rates, and intrastate energy transmission and transportation. America's federalism may be "cooperative" at many key points of national-state contact, but it is certainly not uniform in its standards and requirements. Indeed, state variations quite often frustrate the intent of national lawmakers. The result for national energy and environmental policy is an even more highly refined fragmentation.

Pork-barrel "Distributive" Politics

Historically, the United States energy base has undergone fundamental shifts, from wood to coal, to oil and natural gas (both for direct power and electrical power generation), and now, presumably, to atomic fuels. Regardless of these major transitions in energy sources, at least one form of political process governing the allocation, subsidization, and access rights to these resources has, until very recently, remained largely unchanged. That is the classic pork-barrel model of American politics, which various political scientists have dignified with the title "distributive politics." This form of political activity usually requires little introduction, even for those who are normally inattentive to American politics. Typical popular images brought to mind by the very mention of "pork-barrel politics"—"logrolling" and "horse trading" as politicians "cut up the public pie"—illustrate this form of activity.

The central idea of distributive politics is that there are government agents, both executive and congressional, who have the authority to disburse public moneys for special projects or to dispense other goods held in public trust—land, water, resource access rights, and the like. During the wood-power age, for example, federal lands were sold for a song and given away wholesale via the states. Through a succession of legislative acts—preemption sales, homestead allowances, timber-culture grants, and railroad land grants, just to name a few—the federal government distributed all but some 29 percent of its originally acquired public domain. Improper and outright corrupt administration of the remainder by the old Public Land Office decimated much of what was left. In more recent years, this pattern was extended to the leasing of subsurface mineral rights, franchising of public utility companies, licensing of nuclear power plants, and other power- or environment-related public projects, though probably with a greater sensitivity to public needs than in the past.

A critical feature of distributive politics often missed by observers concerns the understanding and perception of political actors involved. Distributive politics presumes that the stakes at issue—the public goods, rights, or subsidies to be given away or sold—are unlimited in character. Of course, all such commodities have finite limits, but in an earlier era these limits were far more difficult to perceive or at least to take seriously. Consider, for example, America's first power resource, wood. In spite of its singular importance as a fuel and building material, timber waste was staggering, probably over two-thirds of what was cut. The prevailing understanding of early-nineteenth-century Americans was simply that the forest was endless, with more than enough wood for everyone.

A perception of this sort quite naturally led to both the physical and political pattern of distribution; first of the public domain and, later, public-works projects. Until recently, in fact, the possibilities for new public-works projects—canal digging, harbor dredging, levy construction, airport location, power-plant siting, and so forth—were commonly regarded as unlimited. However, the realization that even these sacred preserves of pork-barrel politics are open to challenge on grounds of declining relative benefits and increasing environmental damages was inescapable as the overall limits and interconnections between natural resources and public projects became more widely understood.

The process and structure of distributive politics held important consequences for national energy policies. In the first place, each specific type of energy resource—woodlands, coal mines, oil and gas, and hydroelectricity—has been treated in a largely exclusive way, set apart from others, and without special consideration for the long-term need to coordinate the vital interconnections among them. In an era when almost all energy sources were perceived to be superabundant, this simply did not appear to be necessary.

The structural result of dealing separately with energy resources, politically speaking, has been the historical development of relatively isolated decision-making units of the sort that Washington journalist Douglass Cater has termed "subgovernments." These centers for specialized energy and resource decision typically revolve around an executive-branch agency, such as the Bureau of Mines, the Office of Oil and Gas in the Department of the Interior, or an independent regulatory commission such as the Federal Power Commission or the Atomic Energy Commission. Within this orbit are equally specialized congressional committees along with nationally organized special-interest groups, such as the American Petroleum Institute and the American Mining Congress. There is no authoritative, overall coordination.

In the second place, expected and actual consequences of this political structure are resource and environmental policies that are unconnected and often contradictory. Each of the specialized clusters of political power—the agency or independent regulatory commission with its associated congressional committees, subcommittees, and private-interest organizations—makes its own energy-resource decisions with scant regard for the others or for an integrated natural-resources and energy program. Private interest and business profits were well served by such a system on the implied assumption that resource use decided on this basis would also serve the common good.

"Pluralist" Energy Politics

Even before the close of the last century, the myth of infinite natural resources came under attack. The Sierra Club had already emerged as a political force to be reckoned with, and numerous high-ranking government officials repeatedly issued warnings about the rapid and nationally visible depletion of forest and soil resources. What is more, the number of competing claims on the public domain had grown tremendously. Of course, at the turn of the century, hydroelectric power, reclamation, mass and motorized-recreation interests, and surface-mining claims, were in the early stages of development or far in the future. Yet, grazing interests were already at loggerheads with timbermen, and newly formed conservation-preservation groups such as the Izaak Walton League and the American Forestry Association were challenging business interests on issues of exclusive and untrammelled use of public lands.

A growing awareness that natural resources had finite limits and the inescapable political fact of new demands for publicly owned resources gradually changed the pork-barrel "subgovernments." This is not to suggest that the old bureaus, congressional committees, and interest groups were swept away in a great burst of new understanding and reform. What did occur was the gradual enlargement of existing decision-making centers with the addition of new claimants on the national bounty. Oil and natural-gas interests, conservationists, and sportsmen were added to the then standard claims of railroad companies, miners, loggers, and ranchers. In short, more types of interests organized themselves nationally to compete for the same limited resources.

In addition, new governing bodies were created to supervise fresh management concepts such as multiple-use and sustained-yield forestry (U.S. Forest Service, 1906), areawide, broad-scale resource administration (Tennessee Valley Authority, 1933), and national air- and water-pollution control (Environmental Protection Agency, 1969). Other administrative and regulatory agencies were set up in response to technological breakthroughs, such as national electrical-power transmission (Federal Power Commission, 1920), the capacity for massive production of hydroelectric power (Bonneville Power Authority, 1937), and atomic power (Atomic Energy Commission, 1944).

These developments were quite important, yet they did not fundamentally alter the basic means of policy-making: legislation, budgeting and appropriations, administrative regulation, and court interpretation. They did change the character of activity in each policy-making forum. Legislation, for example, became a far more competitive process as a greater number and variety of interests contended for a decreasing quantity of available resources. Where private interests had once come singly as entrepreneurs or corporations seeking public franchises or subsidies, now they came coalesced, even in formal alliances. The result was what Bertram Gross has so aptly termed "the legislative struggle." In fact, there were many legislative struggles, as the points of access (government agencies and congressional committees) open to private interests proliferated. State and local authorities offered still additional battlefields for private interests pressing public claims.

Each of the traditional means of public policy-making increasingly had to deal with a plurality of competing private interests. New parties testified before legislative and administrative hearings. New demands were made for funded-program assistance. New litigants appeared in the courts. Policy outcomes gradually changed from straight-forward concessions for private claimants to middle-ground compromises between contending organizations and alliances or, in a word, "pluralism."

Whatever the merits of political pluralism as a surrogate or supplement to representative democracy, the results of pluralist politics for national energy supply and environmental values are fragmented "minipolicies" which frequently have unintended and contradictory side effects. For example, the immense Interstate Highway Program developed during the 1950s and 1960s held fundamental consequences for domestic oil reserves, protection of open land, urban space utilization, and additional research on new energy sources and alternate forms of mass transportation. Each of these effects could have been predicted, their social, economic, and political consequences to some extent calculated, and other possibilities considered. But the Bureau of Public Roads and the congressional subcommittees on highways had no mandate or authority to decide overall transportation policy, let alone comprehensive energy and environmental policy.

A similar situation involves the Tennessee Valley Authority. Created as an independent government corporation, TVA has evolved from its conception as a seven-state, resource-conservation agency and low-cost hydroelectric power producer to stimulate economic development to its present position as a major commercial producer of coal-generated electricity. TVA's board of directors has made this transition over years of careful consultation with its organized (mostly business) constituents and with the obliging consent of its congressional committee overseers. Yet TVA's subsequently voracious demand for strip-mined coal has resulted in severe environmental consequences for the Appalachian region. No national authority existed to weigh the costs and benefits of government electricity production against a vast acreage of unrestored strip-mine wasteland. The U.S. Bureau of Mines and the congressional subcommittees on mines retained an interest in mining safety, of course, but the matter of general strip-mining regulation and environmental controls was left to the states. For their part, state governments largely withdrew from active controversy, relying instead on antiquated mining laws and subsurface rights granted before the invention of strip-mining techniques.

Pluralist policy-making is not only fragmented and specialized, it is also incremental, fashioned by small annual or semiannual additions to an existing policy base. Economist Charles Lindblom has identified this process as the method of "successive limited comparisons," which involves periodic reexamination by a policy-making body and active response only to those data and conditions falling within its specific jurisdiction and relating to its basic framework assumptions. By way of illustration, the Bureau of Mines has not aggressively attempted to take regulatory control over multistate strip-mining operations or to coordinate coal mining with other forms of subsurface energy-resource extraction, fossil-fuel research, or environmental protection. Instead it has taken a more cautious and defensive stance, working incrementally within the narrowly construed boundaries of the bureau's enabling legislation. This approach to policy-making avoids comprehensive reexamination of existing programs or unsettling departures from an incrementally developed past. Such new initiatives regularly require the creation of new agencies and legislative overseers, which of course creates yet additional specialized centers of policy-making power in an even more fragmented political setting.

"Veto" Politics

According to the model and the reality of pluralist politics, the outcome of each struggle for policy advantage is a compromise or a collection of minor compromises. Professor Theodore Lowi has argued that outcomes of this process are better described as "residues" than as policies. Regardless of one's assessment, pluralism is premised on the understanding that contending parties to a dispute can and should compromise. And in the case of environmental and energy-resource politics —in their separate and numerous decision-making centers—the realization that natural resources are limited has not prevented the adjustment of competing claims made by industry, business associations, and other private groups.

Once decision-makers understand that some limited resources are also nonrenewable, that used or destroyed they can never by fully replenished by nature, then there is an increasing probability that pluralist politics will be stalemated. But a stalemated outcome has fundamentally to do with the perceptions of those involved. If environmentalists believe, for example, that a delicately balanced natural area, estuary, or one-of-a-kind natural wonder like the Grand Canyon or the Florida Everglades will be irreversably damaged by the installation of a new power plant or the development of some other facility, then pluralism's compromise answer is no compromise at all.

The American political system has always required a genuine and sustaining legislative majority before any positive program could be initiated and continued. What is more, program proposals have had to pass muster under scrutiny by any number of potential "veto groups" before they could gain the force of law. Congressional standing committees with policy jurisdictions, the House Rules Committee, the House and Senate appropriations committees, executive line agencies, the Office of Management and Budget, the federal courts, coalitions of interest groups—each of these political participants has some form of policy-veto potential, in addition to the formal legislative veto constitutionally vested in the president. Consequently, an "environmental veto" to protect unique or nonrenewable natural resources is a distinct possibility.

Environmental vetoes have actually been used in a number of instances to block the destruction of nonrenewable resources. In 1971 the Cross Florida Barge Canal, the Turkey Point (Biscayne Bay, Florida) Atomic Energy Plant, and the Everglades Jetport were all "vetoed" by presidential action on the ground that they would irreparably damage the environment. Even more recently, the Supreme Court vetoed construction of the Alaska pipeline, at least for so long as it held jurisdiction over the project. State government officials have also begun to exercise "veto" powers to preserve resources as public pressure for land and water protection has become more widespread.

Although the environmental veto may become a relatively significant variant of pluralist politics relevant to national energy policy, several rather severe prerequisites apply if such vetoes are to be effective. In the first place, the veto agent must possess the recognized power to block affirmative action. The president, for example, has used budgetary impoundment as a means to block environmentally damaging projects. If this particular power is successfully challenged by Congress and in the courts, obviously it will no longer be available as a veto tool. Secondly, the vetoing agent must be determined to continue the effect of his veto. Should the resolve of a governor or a congressional committee weaken in an effort, say, to prevent an environmentally threatening nuclear-energy plant, mining operation, or dam, the power of decision will revert to the original pluralist contest. There, so long as the pressures for greater energy production and resource development continue, a compromise-policy outcome will constitute a clear-cut decision against nonrenewable resource conservation. Finally, the political actor attempting to sustain an environmental veto must retain jurisdiction if he is to succeed. A veto decision issued in the strongest terms, such as the Supreme Court's 1973 decision to halt construction of the Alaska pipeline, may be rapidly overturned where jurisdiction is shared with others. In the Alaskan case, Congress changed the legislative basis for the Court's action by exempting the pipeline from provisions of the Environmental Protection Act, pulling the jurisdictional rug out from under the Court entirely.

Environmental vetoes may be critically important to the protection of delicate ecosystems and special scenic areas, but they are hardly the fundament of comprehensive policy. Since such vetoes are controversial in character, flaunting the norm of compromise, they are issued only sporadically, and probably for idiosyncratic reasons as well. Essentially, they are a negative force, blocking positive action, and almost irrelevant to the identification of policy alternatives or the development of environmental and energy policies covering a wide range of considerations. It may be, in fact, that environmental vetoes have greater symbolic than actual value.

In a tug of war between environmental and energy interests, illustrated by the Alaska pipeline dispute or by the many intense struggles over nuclear-power-plant sites, environmental vetoes are presently unlikely to have a long-term effect. The nation's voracious appetite for increased energy may indeed eclipse environmental values where this sort of energy-environment confrontation is made clear. In fact, the negative force of veto actions may well be deployed against environmental values in order to guarantee short-run availability of energy.

Comprehensive Policy

The present pattern of energy and environmental policy-making fragmented as it is by specialized centers of decision-making power, appears quite unlikely to change until the perceptions of policy-makers and their supporting publics are drastically altered. This will not occur until most environmental and energy resources are understood to be not only limited and nonrenewable in character but also thoroughly interrelated. Such an outlook, focused by energy scarcity and environmental damage of crisis proportions, demands a political reordering.

Comprehensive environmental and energy policy-making restructuring is virtually sure to include a reorganization of presently disaggregated legislative and administrative powers around a single set of executive initiatives, guidelines, and vetoes. Such a reordering still leaves room for private lobbying, coalition building, bargaining, and some compromise. No doubt a few juicy morsels will still be found in the pork barrel of potential public-resource projects. But the possibility of clear-cut policy decisions of the "yes" or "no" variety would also be allowed, both to protect critical environmental values and to guarantee vital energy sources. At the same time the stakes at issue would become more apparent to far greater numbers of people and their organizations, thus increasing both the size and the intensity of the energy-environment contest. Yet the structure of decision at the point of genuine crisis is likely to be more centralized, and probably more arbitrary as well, overriding state and local policy variations and congressional idiosyncracies.

The policy outcomes of "comprehensive" politics would presumably be far better integrated, more responsive to national priorities, and more consciously related to foreign-policy considerations. The incremental basis of pluralist decision-making, adding bit by bit to established policy, might be temporarily suspended, but the essential character of competitive group politics could be preserved. Also the aspect of contemporary politics that Murray Edleman has termed "symbolic reassurance" will almost certainly have a continued place in a reordered political setting as policy-makers attempt to reassure the citizenry with proclamations and symbolic acts of concern while genuine solutions to thorny energy and environmental problems remain illusive or politically untenable.

Speculation about a new and much more centralized decision structure for an energy- and environmental-resource policy is not entirely hypothetical. The Nixon administration has moved slowly but deliberately in this direction. Passage of the Environmental Protection Act of 1969, following hard on the heels of tough new federal air- and water-pollution acts, the administration's proposed "Department of Natural Resources," the federal land-use bills of 1971-72, the pending Land Use Policy and Planning Assistance Act of 1973, and the presidential appointment of a national energy coordinator—all of these developments may be regarded as tactics in a relatively consistent effort to centralize natural-resources policy.

Even in the best of times for a presidential administration, the goal of a more centralized and rational policy-making structure is not so easily obtained. Congressional committees are unlikely to be any more receptive to the comprehensive resource-use implications of an Environmental Protection Agency or national land-use planning than they were to the planning-programming-budgeting systems (PPBS) installed in the Department of Defense in 1961 and ordered for all federal departments by President Johnson in 1965. The committees, along with willing agency accomplices, consistently undermined PPBS, as they have now begun to undermine and attack outright provisions of the Environmental Protection Act. The unstated objection is that each of these information systems forces different and more difficult decision-making tasks on Congress. New and conflicting resource-use values must be overtly considered and alternatives directly chosen. In this process pet projects and political rewards are more easily brought into an unflattering light.

Resource-policy centralization also faces the standing challenge of bureaucratic resistance. Cherished agency jurisdictions over policies, long-standing procedures for doing business, and administrative careers may seem threatened by a proposed reorganization of executive authority.

It need hardly be added that the centralized presidency is itself a live issue in the 1970s. Proposals for executive reorganization, which are always controversial, will no doubt be fought with special vigor during the Nixon administration. In addition to the opposition of agencies and congressional committees, reorganization and centralization of land-use, environmental and energy policy face the possibility of an unusual and especially powerful coalition of opponents. Federal land-use ("zoning") acts are quickly viewed as a threat by state and local authorities. State officials and many congressmen resist the intervention of "outsiders" and "national planning." Nationally operating corporations have had the advantage of widely deviating state environmental regulations and special allowances to attract new industry. Thus lobby associations representing these industries in Washington seem unlikely to champion the imposition of national environmental standards. Paradoxically, conservation groups, which have achieved striking gains in the form of statewide land-use and development planning acts in Florida, Delaware, and Vermont, and partial or pending successes in other states, may be reluctant to abandon the promise of aggressive state action. Several national groups have already expressed concern that federal managers, eager to meet national energy demands, may be more permissive of resource exploitation than their state counterparts. Furthermore, a long-standing belief in the sanctity of private land from federal dictates remains fundamental to American political thinking. Indeed, this factor alone offers a formidable impediment to major changes in American land-use and natural-resources policy.

The political barriers to a national energy and environmental policy are thus numerous and substantial. The extended history of distributive pork-barrel politics has given the nation a tradition of specialized and fragmented natural-resource decisions, made piecemeal and without much thought to overall effects. The contemporary variants of this tradition—political pluralism with occasional environmental vetoes—have provided for increased competition over the use of scarce resources, but the structural fragmentation of energy and environmental policy-making persists. What is more, primary participants in the political process—agencies, congressional committees, and private-interest associations—are likely to find far greater costs than benefits in any structural change in the direction of comprehensive natural-resource decision-making. In the present political setting, each of these participants is situated to make a determined political fight against such change. Not the least of barriers to change is the matter of natural-resource perception. A heritage of almost unlimited national abundance of energy and other natural resources and the accustomed free-wheeling entrepreneural use of these resources, abetted by the political system, is not rapidly overcome, even with the tools of modern communications and persuasion. Political barriers of this magnitude are not likely to be broken down until combined energy and environmental problems reach emergency levels.